Advances in
CANCER
RESEARCH

Volume 106

Advances in
CANCER RESEARCH

Volume 106

Edited by

George F. Vande Woude
Van Andel Research Institute
Grand Rapids
Michigan, USA

George Klein
Microbiology and Tumor Biology Center
Karolinska Institute
Stockholm, Sweden

AMSTERDAM • BOSTON • HEIDELBERG • LONDON
NEW YORK • OXFORD • PARIS • SAN DIEGO
SAN FRANCISCO • SINGAPORE • SYDNEY • TOKYO
Academic Press is an imprint of Elsevier

Academic Press is an imprint of Elsevier
525 B Street, Suite 1900, San Diego, CA 92101-4495, USA
30 Corporate Drive, Suite 400, Burlington, MA 01803, USA
32 Jamestown Road, London, NW1 7BY, UK
Linacre House, Jordan Hill, Oxford OX2 8DP, UK
Radarweg 29, PO Box 211, 1000 AE Amsterdam, The Netherlands

First edition 2010

Copyright © 2010 Elsevier Inc. All rights reserved.

No part of this publication may be reproduced, stored in a retrieval system
or transmitted in any form or by any means electronic, mechanical, photocopying,
recording or otherwise without the prior written permission of the Publisher.

Permissions may be sought directly from Elsevier's Science & Technology Rights
Department in Oxford, UK: phone (+44) (0) 1865 843830; fax (+44) (0) 1865 853333;
email: permissions@elsevier.com. Alternatively you can submit your request online by
visiting the Elsevier website at http://elsevier.com/locate/permissions, and selecting
Obtaining permission to use Elsevier material.

Notice
No responsibility is assumed by the publisher for any injury and/or damage to persons
or property as a matter of products liability, negligence or otherwise, or from any use
or operation of any methods, products, instructions or ideas contained in the material
herein. Because of rapid advances in the medical sciences, in particular, independent
verification of diagnoses and drug dosages should be made.

ISBN: 978-0-12-374771-6
ISSN: 0065-230X

For information on all Academic Press publications
visit our website at www.elsevierdirect.com

Printed and bound in USA
10 11 12 10 9 8 7 6 5 4 3 2 1

Working together to grow
libraries in developing countries

www.elsevier.com | www.bookaid.org | www.sabre.org

ELSEVIER   BOOK AID International   Sabre Foundation

Contents

Contributors to Volume 106   vii

Update on Human Polyomaviruses and Cancer
Ole Gjoerup and Yuan Chang

   I. Introduction   1
  II. Biology of Polyomaviruses   3
 III. Merkel Cell Polyomavirus and Human Cancer   10
  IV. SV40 as a Classic Model System   16
   V. Conclusion   34
      References   34

The Tyrosine Phosphatase Shp2 in Development and Cancer
Katja S. Grossmann, Marta Rosário, Carmen Birchmeier, and Walter Birchmeier

   I. Introduction   54
  II. Shp2 Activation and Signaling   54
 III. Lessons from Human and Murine Gain-of-Function Mutations of Shp2   59
  IV. Lessons from Loss-of-Function Mutations of Shp2   63
   V. Organ Specific Roles of Shp2: Lessons from Mutations in Mice   67
  VI. Conclusions and Perspectives   77
      References   78

CXC Chemokines in Cancer Angiogenesis and Metastases
Ellen C. Keeley, Borna Mehrad, and Robert M. Strieter

   I. Introduction   92
  II. Angiogenic CXC Chemokines and Receptors   93
 III. Angiostatic CXC Chemokines and Receptors   93
  IV. Immunoangiostasis   95
   V. Chemokine-Induced Angiogenesis in Tumor Models   96

VI.	Chemokines Affect on Cancer Metastases 102
VII.	Conclusion 104
	References 104

Genetically Engineered Mouse Models in Cancer Research
Jessica C. Walrath, Jessica J. Hawes, Terry Van Dyke, and Karlyne M. Reilly

I.	Introduction 114
II.	Generation of Genetically Engineered Mouse Models 116
III.	Cancer Paradigms and Lessons from the Mouse 141
IV.	Summary 153
	References 154

Index 165

Color Plate Section at the end of the book

Contributors

Numbers in parentheses indicate the pages on which the authors' contributions begin.

Carmen Birchmeier, Neuroscience Department, Max-Delbrueck-Center for Molecular Medicine, Berlin, Germany (53)

Walter Birchmeier, Department of Cancer Research, Max-Delbrueck-Center for Molecular Medicine, Berlin, Germany (53)

Yuan Chang, Cancer Virology Program, Hillman Cancer Research Pavilion, University of Pittsburgh Cancer Institute, Pittsburgh, PA, USA (1)

Ole Gjoerup, Cancer Virology Program, Hillman Cancer Research Pavilion, University of Pittsburgh Cancer Institute, Pittsburgh, PA, USA (1)

Katja S. Grossmann, Department of Cancer Research; and Neuroscience Department, Max-Delbrueck-Center for Molecular Medicine, Berlin, Germany (53)

Jessica J. Hawes, Mouse Cancer Genetics Program, National Cancer Institute, Frederick, Maryland, USA (113)

Ellen C. Keeley, Division of Cardiology, Department of Medicine, University of Virginia, Charlottesville, Virginia, USA (91)

Borna Mehrad, Division of Pulmonary and Critical Care Medicine, Department of Medicine, University of Virginia, Charlottesville, Virginia, USA (91)

Karlyne M. Reilly, Mouse Cancer Genetics Program, National Cancer Institute, Frederick, Maryland, USA (113)

Marta Rosário, Department of Cancer Research, Max-Delbrueck-Center for Molecular Medicine, Berlin, Germany (53)

Robert M. Strieter, Division of Pulmonary and Critical Care Medicine, Department of Medicine, University of Virginia, Charlottesville, Virginia, USA (91)

Terry Van Dyke, Mouse Cancer Genetics Program, National Cancer Institute, Frederick, Maryland, USA (113)

Jessica C. Walrath, Mouse Cancer Genetics Program, National Cancer Institute, Frederick, Maryland, USA (113)

Update on Human Polyomaviruses and Cancer

Ole Gjoerup and Yuan Chang

Cancer Virology Program, Hillman Cancer Research Pavilion, University of Pittsburgh Cancer Institute, Pittsburgh, PA, USA

I. Introduction
II. Biology of Polyomaviruses
   A. Classification and Phylogeny
   B. Genome Organization
   C. Viral Life Cycle
   D. Natural Infection, Reactivation, and Clinical Disease
III. Merkel Cell Polyomavirus and Human Cancer
   A. Human Polyomaviruses and Cancer
   B. Merkel Cell Polyomavirus
IV. SV40 as a Classic Model System
   A. Large T Antigen
   B. Small t Antigen
   C. Transgenic Model Systems
V. Conclusion
   References

Over 50 years of polyomavirus research has produced a wealth of insights into not only general biologic processes in mammalian cells, but also, how conditions can be altered and signaling systems tweaked to produce transformation phenotypes. In the past few years three new members (KIV, WUV, and MCV) have joined two previously known (JCV and BKV) human polyomaviruses. In this review, we present updated information on general virologic features of these polyomaviruses in their natural host, concentrating on the association of MCV with human Merkel cell carcinoma. We further present a discussion on advances made in SV40 as the prototypic model, which has and will continue to inform our understanding about viruses and cancer. © 2010 Elsevier Inc.

I. INTRODUCTION

Five human polyomaviruses have been identified to date. In 1971, the first two members, JC virus (JCV) and BK virus (BKV) were concurrently reported in the journal Lancet (Gardner *et al.*, 1971; Padgett *et al.*, 1971).

JCV was cultured from progressive multifocal leukoencephalopathy (PML) brain tissue in a patient with Hodgkin disease and BKV was isolated from the urine of a renal transplant patient with ureteral stenosis. Both viruses manifested unusual clinical diseases in immunosuppressed patients and were named after their source patients' initials. More than 30 years intervened before two more human polyomaviruses were identified in 2007 (Allander *et al.*, 2007; Gaynor *et al.*, 2007). These two viruses, Karolinska Institute virus (KIV) and Washington University virus (WUV) were named after the institutions where their identification occurred. Both were detected from respiratory samples in symptomatic pediatric patients after DNase enrichment for encapsidated viral particles followed by library construction and mass sequencing of cloned cDNAs. The publication of the most recently identified human polyomavirus, Merkel cell polyomavirus (MCV) occurred in 2008 (Feng *et al.*, 2008). MCV was named for the uncommon, but aggressive Merkel cell carcinoma (MCC) skin cancer, from which viral transcripts were found by digital transcriptome subtraction (DTS) (Feng *et al.*, 2007). DTS involves transcriptome sequencing followed by *in silico* subtraction to exclude human from candidate viral transcripts (Feng *et al.*, 2007). In contrast to other human polyomaviruses, the discovery of MCV did not depend on the presence of replication competent, encapsidated virions.

Research on polyomaviruses began in 1953 with strong ties to cancer biology when Ludwik Gross, during the course of his investigations in transmitting mouse leukemia from cell-free filtrates, isolated an agent that induced tumors in newborn mice (Gross, 1953). This filterable agent, murine polyomavirus (MPyV) became the archetypal member of the *Polyomaviridae* family. In 1960, Sweet and Hilleman found simian vacuolating virus 40 (SV40) infecting lots of rhesus monkey kidney cells used for the production of both Sabin and Salk polio vaccines (Sweet and Hilleman, 1960). The ability of SV40 to cause tumor in experimental animals and the widespread administration of the polio vaccine raised concerns regarding xenotropic exposure in the human population to this agent. The resultant search over the past 50 years to establish a causal relationship between SV40 and human cancers has not been fruitful; however, this does not diminish the contributions that studies of SV40 have made to both viral oncogenesis and cellular biology. In this review, we will examine the biology of the human polyomaviruses concentrating on MCV and its association with human cancer. SV40, and its gene products, serve as well-established models for cancer and provide a context for these discussions due to commonalities in genome, structure, and biochemical properties with the human polyomaviruses.

II. BIOLOGY OF POLYOMAVIRUSES

A. Classification and Phylogeny

Polyomaviruses were historically categorized with the papillomaviruses under the designation of papovaviruses until their separation into two distinct families in 2000. In addition to MPyV, SV40, and the five human members, other polyomaviruses from diverse species of animals including other nonhuman primates, birds, bats, rabbits, and rodents have been found. To date, full genome sequences of over 21 polyomaviruses have been deposited in GenBank. Although variations exist with respect to phylogenetic relatedness when different genes are used for analysis, JCV, BKV, KIV, and WUV appear to group closely with SV40 (Fig. 1). Within this subgroup JCV and BKV are consistently less divergent from each other when compared to either KIV or WUV, which appear to be in a clade of their own. In contrast to the first four identified human polyomaviruses, MCV is more closely related to the archetypal murine polyomavirus (MPyV) and the African green monkey lymphotropic polyomavirus (LPV) (Fig. 1).

B. Genome Organization

The genomes of human polyomaviruses range between $\sim$5.0 and 5.3 kb (JCV-5130 bp; BKV-5153 bp; KIV-5040 bp; WUV-5229 bp; MCV-5387 bp). They exist as circular dsDNA closely associated with histones, and are packaged into chromatin resembling cellular genomes (minichromosomes) within nonenveloped, 40–45 nm icosahedral capsids. The polyomavirus genome is almost evenly divided into an early and a late region encoded on opposite strands. These two regions are separated by a noncoding regulatory region (NCRR) containing the origin of replication and transcriptional control elements. The SV40 genome is depicted in Fig. 2, since we will later use it as a model for polyoma-induced neoplastic transformation. The early region is transcribed from the early promoter immediately upon entry and uncoating of the genome, while the late region is expressed from the late promoter after the onset of viral DNA replication. Early message is differentially spliced to encode at least two proteins, and up to five in some polyomaviruses. The large T (tumor) antigen (LT) and small t antigen (ST) proteins are invariantly expressed in all polyomaviruses including the five human members, although in KIV and WUV these two proteins have been predicted only by open reading frame analysis and not by experimentation (Fig. 3). These early proteins are referred to as tumor antigens, because they were originally detected using antibodies from tumor bearing animals.

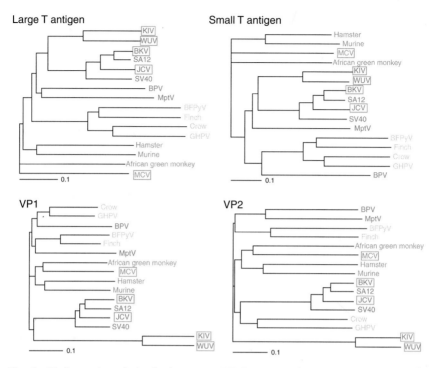

Fig. 1 *Phylogenetic analysis of polyomavirus LT, ST, VP1, and VP2 protein sequences.* The analysis includes: human polyomaviruses (BKV, JCV, KIV, WUV, and MCV, all marked with a green rectangle) as well as simian agent 12 (SA12), SV40, bovine polyomavirus (BPV), murine pneumotropic virus (MptV, also known as Kilham strain of polyomavirus), budgerigar fledgling disease polyomavirus (BFPyV), finch polyomavirus, crow polyomavirus, goose hemorrhagic polyomavirus (GHPV), hamster polyomavirus, murine polyomavirus, African green monkey polyomavirus (also known as lymphotropic polyomavirus), and MCV. The subgroup of SV40 is outlined in blue, of the murine polyomavirus in red, and of the avian polyomavirus in orange. While BKV, JCV, KIV and WUV LT and ST sequences cluster together with SV40, MCV, in contrast, clusters with the murine polyomavirus subgroup (modified from Feng *et al.*, 2008; Fig. 2B). (See Color Insert.)

Analogous to the 17k T antigen of SV40 which is expressed from an alternatively spliced early transcript (Zerrahn *et al.*, 1993), additional T antigen isoforms have also been identified in JCV, BKV, and MCV: JCV encodes three T′ antigens, T′165, T′136, and T′135 (Trowbridge and Frisque, 1995); MCV has a 57 kDa T antigen (57kT) (Shuda *et al.*, 2008) (Fig. 3); and BKV encodes a truncated T antigen close in structure to T′136 of JCV (Abend *et al.*, 2009b). No accessory T antigens have yet been identified for KI or WU. Regardless of the number of T antigen mRNAs, exon 1 is shared in common with all alternatively spliced early mRNAs of each virus (Fig. 3). The functions of the accessory T antigen proteins are still largely unknown. Determination of their function in the viral life cycle awaits

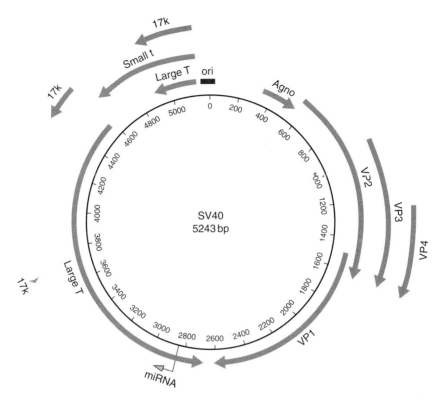

Fig. 2 *SV40 genome organization*. The early region of the viral genome (the left half) encodes LT, ST, and 17k by differential splicing. The respective open reading frames are colored blue. The late region of the viral genome (the right half) encodes agnoprotein and the structural proteins VP1, VP2, VP3, and VP4. These gene products are generated by differential splicing and internal translation. The core origin of replication (ori) is located on top adjacent to transcriptional control elements, together encompassing the NCRR. A red arrow indicates the viral miRNA that targets the early message. (See Color Insert.)

generation of mutant viruses lacking their expression. It is known that the 17k SV40 accessory T is expressed at low levels during SV40 infection, where it has been suggested to fine-tune cell-cycle regulation (Zerrahn *et al.*, 1993). It can bind pRB family members, drive cell-cycle progression, and is capable of causing minimal transformation of F111 rat cells (Boyapati *et al.*, 2003; Zerrahn *et al.*, 1993). A mutant 17k deficient in pRB binding drives normal human fibroblasts into premature senescence (Gjoerup *et al.*, 2007).

The late message, by differential splicing and internal translation, produces three to four capsid proteins VP1, VP2, VP3, and VP4. VP3, and VP4 when present, is generated by internal translation of VP2. VP4 has so far only been detected in SV40, where it promotes lysis of the cell and egress of

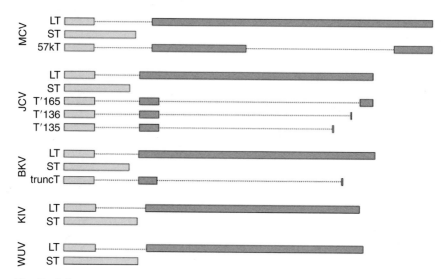

Fig. 3 *Splicing arrangement for the human polyomavirus T antigens.* Colored rectangles indicate coding sequences, whereas the broken line depicts intron sequences. Different color rectangles refer to distinct reading frames after splicing events. All T antigens of each polyomavirus share the sequence encoded within exon 1. The accessory T antigens in addition share fragments of their respective LT sequences. (See Color Insert.)

the virus (Daniels *et al.*, 2007). The viral capsid is composed of 72 pentamers of the major capsid protein VP1 that contacts 72 copies of the minor capsid proteins VP2/3. In BKV and JCV, the leader region of the late message additionally encodes the agnoprotein, which may be involved in virion maturation (Khalili *et al.*, 2005; Ng *et al.*, 1985). KIV, WUV, and MCV are without known agnoproteins. Only murine and hamster polyomaviruses are known to encode a middle T antigen, which is their principal transforming protein (Cheng *et al.*, 2009).

Several polyomaviruses have been found to express miRNAs derived from the late transcript. SV40, MPyV, JCV, and BKV each has a single pre-miRNA from which two miRNAs of opposing orientations are processed (Seo *et al.*, 2008; Sullivan *et al.*, 2005, 2009). The JCV miRNAs have been detected in PML lesions (Seo *et al.*, 2008). These miRNAs are predicted to autoregulate early gene expression at late times in infection. For SV40, mutants that cannot produce miRNA show increased expression of viral T antigens and are more susceptible than wild-type virus to lysis by cytotoxic T cells (Sullivan *et al.*, 2005). In contrast, no *in vivo* differences in antiviral CD8 T cell responses can be discerned between infections with wild-type MPyV and MPyV lacking miRNA (Sullivan *et al.*, 2009).

C. Viral Life Cycle

The polyomaviruses generally have a narrow host range and limited cell type tropism. They are able to infect cells of their natural hosts, giving rise to a productive life cycle that results in cell lysis. In addition, these viruses are also able to establish a persistent infection rarely associated with disease except when immunodeficiency is induced. The full, infectious viral life cycle has only formally been studied in JCV and BKV as no infectious system has yet been devised for KIV, WUV, or MCV. However, it is generally thought that polyomaviruses are internalized by the interaction of VP1 with specific cellular receptors and co-receptors. It is known that BKV uses gangliosides GD1b and GT1b (Low *et al.*, 2006); JCV uses GT1b and the serotonin receptor, 5HT2AR (Elphick *et al.*, 2004); and GT1b has been proposed as a putative host cell receptor for MCV (Erickson *et al.*, 2009). Virus then traffics through caveolae and the endoplasmic reticulum to the nucleus, where it is uncoated and the early message is transcribed. In the case of JCV, clathrin-dependent endocytosis precedes localization in caveosomes (Eash *et al.*, 2006). After translation, LT then initiates DNA replication of the viral genome from the origin in the NCRR. The shift to late viral expression is not fully elucidated but likely involves LT transcriptional activation of the late and repression of the early promoter. VP1 is expressed and assembled together with VP2 and VP3 and then imported into the nucleus where encapsidation of viral genomes takes place.

D. Natural Infection, Reactivation, and Clinical Disease

All five human polyomaviruses have high prevalence in the human population and infections start in childhood; however, variability has been reported in the age patterns of polyomavirus acquisition. A study of 2435 sera from English and Welsh individuals ranging from 1 to 69 years showed an overall seroprevalence rate of 81% for BKV with seroconversion occurring very early in childhood, peaking at 91% in the 5–9 age range, and dampening in elderly individuals. For JCV, the overall seroprevalence was 35% with a steady increase of 11% from children below 5 years up to 50% in the 60–69 age group (Knowles *et al.*, 2003). These findings were largely replicated in a study examining 400 consecutive healthy blood donors from Basel, Switzerland. Egli and colleagues found an overall IgG seroprevalence of 82% for BKV and 58% for JCV in this cohort (Egli *et al.*, 2009). The first large serosurvey of KIV and WUV in 1501 adults using GST-VP1 capture ELISA showed rates of 55% and 69%, respectively (Kean *et al.*, 2009). Age-specific prevalence studies have not been performed for KIV and WUV;

however, the detection of viral DNA in respiratory specimens from children suggests that initial exposure occurs at a relatively early age. Similar findings are emerging for MCV. Using wild-type MCV strain sequences in VLP-based ELISA, Tolstov and colleagues reported an age associated increase in MCV prevalence from 50% among children 15 years or younger up to 80% among persons older than 50 years (Tolstov *et al.*, 2009). Pastrana and colleagues found 88% MCV seropositivity in adults without MCC. These studies additionally demonstrate that MCC patient sera showed markedly elevated MCV IgG responses with the geometric mean titers in controls 59-fold lower than in the MCC patient group (Pastrana *et al.*, 2009; Tolstov *et al.*, 2009). Carter and colleagues reported the following seroprevalence results in 451 general population adults: 92% for BKV, 45% for JCV, 90% for KIV, 98% for WUV, and 59% for MCV (Carter *et al.*, 2009). Although rates of infection for JCV and BKV found in various studies are in general agreement, additional studies on KIV, WUV, and MCV will more precisely define overall and age-specific prevalence rates.

Fecal–oral, oral, and respiratory routes of transmission have been proposed for different human polyomaviruses. For JCV and BKV, studies of urban sewage samples and rivers show significant numbers of stable virus particles from divergent geographical areas suggesting the possibility of virus acquisition through fecally contaminated water, food, and fomites (Bofill-Mas and Girones, 2003; Bofill-Mas *et al.*, 2000; McQuaig *et al.*, 2006). Initial studies of KIV, WUV, and MCV have also shown viral DNA in stool samples (Allander *et al.*, 2007; Babakir-Mina *et al.*, 2009a; Loyo *et al.*, 2009). The detection of salivary shedding of BKV and productive infection of salivary cell lines implicates oral transmission as another possibility for this virus (Jeffers *et al.*, 2009). MCV is also detected at high level in saliva (Loyo *et al.*, 2009). In contrast to JCV and BKV, which are only rarely detected from respiratory sources, KIV and WUV can be isolated from respiratory secretions of children worldwide. Curiously, these viruses cannot be detected in the respiratory specimens of adults except in the setting of immunosuppression (Bialasiewicz *et al.*, 2009; Loyo *et al.*, 2009; Norja *et al.*, 2007; Ren *et al.*, 2008). MCV DNA can also be detected in respiratory specimens from symptomatic patients and occurs in nasal swabs and nasopharyngeal aspirates at a frequency similar to or even higher than that of KIV and WUV (Bialasiewicz *et al.*, 2009; Kantola *et al.*, 2009). Significantly more adults than children are positive for respiratory MCV in contrast to KIV and WUV (Goh *et al.*, 2009).

The process by which polyomaviruses gain access to and establish persistent infections in distal body compartments is not well established. However, DNA of all human polyomaviruses has been detected in tonsillar tissue, a possible point of entry (Babakir-Mina *et al.*, 2009b; Kantola *et al.*, 2009; Monaco *et al.*, 1998a). For JCV, virus has been localized to tonsillar stromal cells and B lymphocytes with the latter cell type implicated in circulatory dissemination to other anatomic sites (Monaco et al., 1998a,b). Since all viruses can be detected at increased frequencies in blood and lymphoid tissues during host immunosuppression (Sharp et al., 2009), it is likely that hematolymphoid cells can carry or harbor polyomaviruses. Recently, Mertz and colleagues detected MCV in (CD14+/CD16−) inflammatory monocytes but not in lymphocytes or granulocytes (Mertz et al., 2009).

JCV and BKV establish persistent infections in renal tissue and virus is shed into the urine. In a recent study examining 400 consecutive healthy blood donors from Basel, Switzerland, Egli and colleagues found urinary shedding of BKV in 7% of these individuals compared to 19% for JCV (Egli et al., 2009). Reactivation of JCV and BKV, as reflected by increased viruria, occurs during immunosuppression, but only BKV levels correlate with the degree of immunosuppression (Behzad-Behbahani et al., 2004). The bone marrow is another possible site of persistent infection for JCV and may be the source of virus positive cells detected in the circulation (Tan et al., 2009). JCV can additionally gain access to glial cells of the central nervous system where reactivation or new infection can result in PML and virus can be detected in the cerebral spinal fluid (Drews et al., 2000; Vago et al., 1996).

PML is an acquired demyelinating disease pathologically characterized by a triad of findings: 1) oligodendrocytes having enlarged nuclei with viral inclusion bodies, 2) bizarre, atypical astrocytes, and 3) loss of myelin with accompanying phagocytic infiltrate. PML underscores how critical the host cell environment is in determining the outcome of a viral infection. Although both oligodendroglial and astroglial cells of the central nervous system are infected, only the oligodendrocytic myelin forming cells support full lytic replication of the virus causing the hallmark loss of myelin seen in PML. By contrast, infection of the astroglial population results predominantly in changes in nuclear morphology, size, and ploidy. The bizarre, atypical astrocytes which are visually indistinguishable from tumor cells in high-grade glial neoplasia may represent an infection by JCV of a cell type that is unable to support a fully lytic viral life cycle. Viral DNA and LT protein expression is detectable in astrocytes but at a much lower frequency and quantity. Whether the JCV infected astroglial population is transiently transformed is an intriguing conjecture. There are scattered, but convincing case reports in the literature demonstrating expression of JCV T antigen in tumors of the central nervous system (Krynska et al., 1999; Pina-Oviedo et al., 2006); however, as a group, the glial neoplasms have not been associated in a consistent way with JCV infection. Uncommon cases and case series of JCV association with other human neoplasia have also been published (see review Maginnis and Atwood, 2009). PML has recently been diagnosed in patients taking natalizumab (Tysabri) (Kleinschmidt-DeMasters and Tyler, 2005; Langer-Gould et al., 2005; Van Assche et al., 2005), causing its temporary withdrawal from the market (Major, 2010). Several mechanisms have been proposed for the effect of natalizumab on JCV including restriction of leukocyte trafficking across the blood–brain barrier or direct inhibition of T cell reaction against JCV (Chen *et al.*, 2009).

For BKV, important clinical diseases occur in the posttransplant setting and relate to type of tissue transplanted. Hemorrhagic cystitis, hematuria, and renal impairment are seen with hematopoetic stem cell transplantation (HSCT) (O'Donnell *et al.*, 2009); and BK-associated nephropathy occurs in 2–5% of renal transplant patients with graft loss in nearly half of these cases (Bonvoisin *et al.*, 2008). Viral reactivation during these complications is robust and can be monitored by DNA-based techniques in blood and urine (Bonvoisin *et al.*, 2008; Cimbaluk *et al.*, 2009; O'Donnell *et al.*, 2009).

III. MERKEL CELL POLYOMAVIRUS AND HUMAN CANCER

A. Human Polyomaviruses and Cancer

Of the five human polyomaviruses, only MCV demonstrates a robust correlation with human cancer. Studies showing MCV T antigen interactions with cellular proteins such as pRB, Hsc70, and PP2A have been performed, but demonstration of the biochemical effects of these interactions have not been published. As noted above, JCV and BKV are associated with important nonneoplastic clinical diseases that have significant morbidity and mortality in immunocompromised individuals; but despite their potential to act as transforming viruses in rodent and *in vitro* cell culture models, no consistent association with human cancers have been found. The reader is referred to excellent reviews on the potential role of these viruses in human cancers (see Abend *et al.*, 2009a; Maginnis and Atwood, 2009; White *et al.*, 2005). KIV and WUV have not yet been found to be associated with human disease and although both of these viruses were originally detected in the respiratory samples of symptomatic children, attributing a causal role for them in respiratory diseases is difficult because of the significant rates of co-detection with other respiratory pathogens (Bialasiewicz *et al.*, 2008; Norja *et al.*, 2007).

B. Merkel Cell Polyomavirus

MCC is an uncommon cancer with an overall age adjusted incidence of 0.24 per 100,000 person-years (Agelli and Clegg, 2003). Overall incidence of MCC has increased three fold from 1986–2001 (Hodgson *et al.*, 2005).

There is a slight male predominance and a strong association with whites/ fair-skinned individuals, advanced age, and sun exposure. The 5-year relative survival is 75%, 59%, and 25% for localized, regional, and distant MCC, respectively (Agelli and Clegg, 2003). Unfortunately, most cases of primary MCC are diagnosed when the disease is no longer localized.

MCC is derived from resident Merkel cells of the skin, which along with associated terminal sensory neuritis, comprise the epidermal mechanoreceptors that allow touch discrimination of fine surface textures (Maricich *et al.*, 2009). Historically, Merkel cells have been thought to be of neuroendocrine derivation due to their elaboration of various neurosecretory markers; however they also express the low molecular weight cytokeratin (CK) 20 suggestive of epithelial origin. MCC typically affects the elderly, but studies have shown that it may occur in younger ages and at an increased frequency in immunosuppressed individuals. Engels *et al.* noted that MCC was increased in both AIDS and posttransplantation populations (Engels *et al.*, 2002), a striking epidemiologic feature that suggests strong immunologic surveillance in controlling MCC development. In 2008, Feng and colleagues specifically sought for an infectious agent in MCC using the DTS technique (Feng *et al.*, 2008). Pooling cDNA libraries made from four MCC lesions, a DTS candidate was detected that showed a significant degree of similarity to the LPV LT. The 5387 bp polyoma genome (MCC350 strain) was sequenced by PCR walking using primers designed from the DTS viral transcript.

Initial studies of MCC lesions from 10 patients showed that although 80% contained Southern blot detectable MCV DNA, there exists a minority subset of MCC cases which do not contain MCV. This estimate has held up in subsequent studies from various laboratories and has been generalized to geographically diverse populations (Becker *et al.*, 2009; Duncavage *et al.*, 2009b; Foulongne *et al.*, 2009; Kassem *et al.*, 2008; Sihto *et al.*, 2009; Touze *et al.*, 2009; Varga *et al.*, 2009). An exception is the finding of only 24% correlation between virus genome and MCC in Australian populations (Garneski *et al.*, 2009). These findings suggest that MCC is a heterogenous disease having at least two pathoetiologies. Pathological and immunophenotypic studies of virus positive MCC cases compared to virus negative cases demonstrate no other markers that can distinguish between these two types of MCC (Shuda *et al.*, 2009). However, data are beginning to accumulate on clinical correlates for MCV positive MCC compared to virus negative cases. Patients with MCV positive MCC tumors had better overall survival than those with MCV negative tumors (5-year survival: 45% vs. 13%, respectively) (Andres *et al.*, 2009b; Sihto *et al.*, 2009). These findings are preliminary since the uncommon nature of these tumors precludes easy assessment of clinical outcomes and treatment responses.

Determination of viral load in MCC lesions by quantitative PCR shows the following average copy number per tumor cell: 5.2 (range 0.8–14.3)

(Shuda et al., 2009) and 10 (range 0.05–173) (Loyo et al., 2009). In contrast, although MCV DNA can be detected in many non-MCC tissues including appendix, colon, lung, peripheral blood mononuclear cells, and skin, the calculated copy number per cell equivalent of DNA is orders of magnitude lower (Loyo et al., 2009; Shuda et al., 2009). Associations have been sought for MCV and other human neoplasia: neuroendocrine tumors from a variety of anatomic sites (Duncavage et al., 2009a; Touze et al., 2009; Wetzels et al., 2009), colon cancer (Kassem et al., 2009), prostate cancer (Bluemn et al., 2009), mesotheliomas (Bhatia et al., 2010), hematopoietic neoplasms (Shuda et al., 2009), breast cancer, ovarian cancer, and skin cancers including basal cell carcinomas, melanomas, and Kaposi's sarcoma (Andres et al., 2009a; Katano et al., 2009; Sastre-Garau et al., 2009; Varga et al., 2009). Results suggest that MCV is specifically associated only with MCC.

Using 3′ RACE, Feng and colleagues detected T antigen–human fusion transcripts in one of the cases of MCC they studied which mapped to a viral integration site in a receptor tyrosine phosphatase, type G (PTPRG) gene intron on Chromosome 3 of the host genome (Feng et al., 2008). The banding pattern on Southern blots indicated monoclonal MCV integration in six of eight MCV-positive MCCs, suggesting that infection and subsequent integration preceded clonal expansion of tumor cells, consistent with MCV being causally linked to MCC. For one MCC with a monoclonal MCV integration pattern, both primary and metastatic tumor samples were available. These two samples exhibited an identical integration pattern, indicating that MCV integration occurred before tumor metastasis (Feng et al., 2008). Sastre-Garau also found clonal integration in all 10 MCV positive MCC tumors they examined. The restriction patterns and integration sites when identified varied between different MCC tumors suggesting that this event is likely to occur at random sites in the host genome (Feng et al., 2007; Sastre-Garau et al., 2009). Sastre-Garau found viral integration in the 3′ end of the T antigen, in the regulatory control region and in the VP1 open reading frame of informative cases (Sastre-Garau et al., 2009).

In nearly all tumor-derived MCV genomes sequenced, missense mutations or deletions in the early region result in the expression of truncated LT antigens (Sastre-Garau et al., 2009; Shuda et al., 2009). These LTs all retain their pocket protein binding motifs, but lose their helicase domains (see below for explanation of functions). By contrast, sequences from the early regions of MCV strains derived from non-MCC tissues show intact T antigen ORF (Shuda et al., 2009). Shuda and colleagues have shown that both tumor- and nontumor-derived MCV LT have LxCxE and DnaJ domains capable of binding pRB and Hsc70 proteins (Shuda et al., 2008). The functional consequence of the tumor-derived T antigen mutations is the retention of pocket protein interactions and the loss of viral DNA replication capability. This is important, because the loss of replication activity upon integration in MCC demonstrates that MCV is not simply a "passenger virus" that happens to grow well in MCC cells. None of these mutations affect the expression of an intact ST (Sastre-Garau *et al.*, 2009; Shuda *et al.*, 2009) which contributes to malignant transformation in the case of SV40 (see Section IV.B). Clearly, adventitious firing from the origin of an integrated viral genome would activate signaling pathways leading to cell death and inhibit tumor outgrowth. The presence of these mutations support the tenet in polyomaviruses biology that lytic replication is incompatible with tumor formation.

In early studies of SV40 transforming potential, it was determined that efficiency of transformation of human fibroblasts was much higher using artificially produced origin defective SV40 mutants and that only defective SV40 could be isolated from transformed monkey, human, and hamster cells (Huebner *et al.*, 1975; Small *et al.*, 1982). For MCV, viral replication has been mapped to a 71 bp core origin (Kwun *et al.*, 2009). All but one of the core origins examined showed invariant wild-type sequences (Kwun *et al.*, 2009). It is possible that overlapping early transcriptional control functions in this region need to be maintained for robust T antigen expression in MCC tumors and alternative mechanisms, such as truncation of C-terminus of T antigen or viral capsid mutations would be required to ablate replication.

PCR detection used in studies to establish an association between various polyomaviruses and human cancers suffers from the exquisite sensitivity as well as susceptibility for template contamination of the technique. A monoclonal antibody, CM2B4, has been developed based on a peptide sequence in exon 2 of MCV LT that also detects 57kT but not ST. The epitope site of CM2B4 is in a unique span of exon 2 not present in other human polyomaviruses (Fig. 4), and therefore not cross-reactive with their T antigens (Shuda *et al.*, 2009). Using this antibody, a high percentage of MCC lesions with PCR detectable MCV DNA show expression of MCV LT antigen localized diffusely to the nuclei of tumor cells, but not in adjacent normal cells (Busam *et al.*, 2009; Shuda *et al.*, 2009). The discrepancy between the few cases which are MCV DNA positive by PCR and T antigen negative by CM2B4 immunoreactivity may be due to very low amounts of T antigen expression, mutational loss of the CM2B4 epitope, or false positive PCR results.

Although many new insights have been gained about MCV, many questions remain outstanding about its biology and the molecular mechanisms whereby it facilitates oncogenesis. Recent evidence indicates that MCV may be widespread in the human body, which raises the question: Are Merkel cells particularly susceptible to transformation by MCV? Alternatively, this tropism might be restricted by transcriptional control from enhancer sequences. Thus far we have substantial evidence that MCV can be found integrated in MCCs, but there is also some indication that it can be found

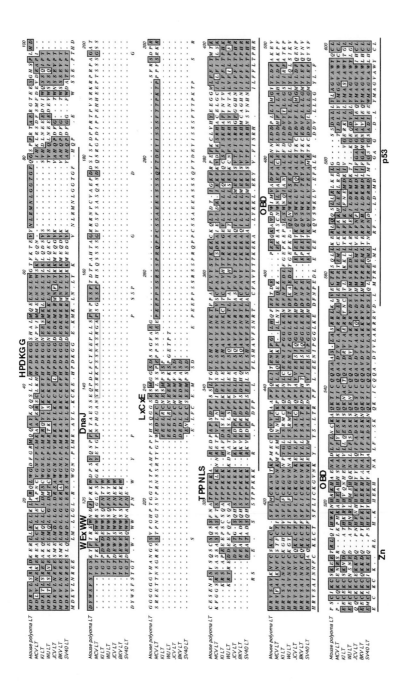

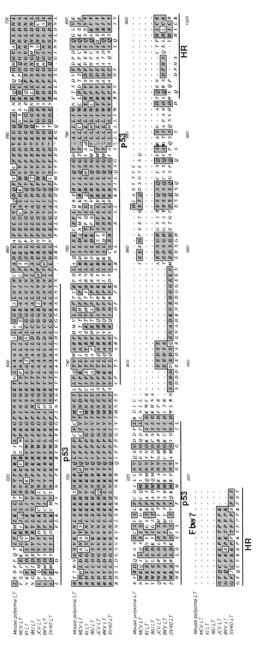

Fig. 4 *Alignment of polyomavirus large T antigens.* The protein sequences of the human polyomavirus, SV40, and MPyV LTs are aligned for comparison. Known domains and motifs are depicted. The first ~70 amino acids constitute the DnaJ domain with the embedded canonical "HPDKGG" motif. The WExWW sequence is a conserved motif required for binding Bub1 in SV40 LT. A red rectangle illustrates the epitope for the CM2B4 MCV LT monoclonal antibody. An intact LxCxE motif is necessary for binding pRB family members. Note that the program (MacVector 10.6.0) did not correctly align the mouse polyoma LxCxE sequence with the others. Both mouse polyoma and MCV LT contain large inserts in their sequence shortly after the first exon. The sequence "TPP" is a conserved phosphorylation site motif, which is proceeded by a canonical nuclear localization signal (NLS). The OBD line indicates the boundaries of the conserved origin-binding domain. The line labeled with "Zn" illustrates the location of a Zn^{2+}-coordination element required for hexamer formation. The line labeled "p53" shows the boundaries of the bipartite p53-binding site in SV40. At the extreme C-terminus the boundaries of the host range (HR) region, which allows SV40 to replicate in certain cell types, is outlined. The extreme C-terminus also contains a binding site for the F-box protein Fbw7 via a phosphodegron sequence. (See Color Insert.)

episomally as well. Other tumor viruses such as HPV are known to be maintained episomally at the precancerous stage; however, in most tumors the genome is integrated resulting in loss of replication activity. Is there a form of MCV "latency" and reactivation, or is the viral genome persistently maintained at low levels? Much future work will depend on development of cell systems that allow us to grow the virus.

Likewise, *in vitro* cell culture systems, and eventually transgenic model systems that recapitulate MCV oncogenic transformation will be paramount. Many outstanding questions relate to the activities of T antigens expressed in MCCs. Given the near universality of viral interference with p53 tumor suppressor signaling, it will be of great interest to find out if, or how, it is perturbed by MCV. Sequence comparisons between different polyoma LTs reveal that, similarly to MPyV LT, MCV has a $\sim$200 amino acids insert immediately following the first exon and before the start of the origin-binding domain. This insert region exhibits no significant homology to other polyomavirus T antigens, suggesting it might confer unique functions (Fig. 4). We are now entering a phase of MCV research where causality for MCC is widely accepted, but the mechanistic details of transformation and the search for therapeutic interventions are underway. A comparison to SV40-induced transformation provides an obvious roadmap; however, MCV promises to reveal additional insights into human oncogenesis.

IV. SV40 AS A CLASSIC MODEL SYSTEM

The human genome is extraordinarily complex. In direct studies that compare the differences between normal cells and tumor cells, for example, by microarray analysis, it becomes extremely challenging to identify the relevant changes that drive malignant progression. Viruses, on the other hand have targeted key molecules and pathways, so-called "hubs" in cellular signaling. Polyomaviruses and their gene products have served as model systems for understanding cellular immortalization and oncogenic transformation. These studies have led to important paradigms and fundamental insight into key biological processes over the last 50 years since they were discovered. The size of polyomaviruses makes them highly amenable to genetic manipulations, and these viruses have historically been easy to grow in culture. Mutants can be tested for proficiency in driving viral replication or transformation, or expression of the viral gene products can be carried out in transgenic models to study tissue-specific tumorigenesis. However, tractable culture systems for the most recently discovered new human polyomaviruses have yet to be developed.

Because of their small genome size, polyomaviruses rely heavily on the cellular replication machinery to replicate their genome. Therefore, these viruses reprogram the host cell cycle to induce progression into S-phase from quiescence and thus create a suitable environment for viral replication. Of relevance to cancer biology, the molecules targeted by the virus to promote unscheduled DNA replication or to inhibit innate immune signaling in the setting of viral replication are often the same as those involved in oncogenesis. In this section we will review and discuss the current knowledge of the biochemical properties of LT and ST using SV40 as our model.

A. Large T Antigen

LT is a multifunctional nuclear phosphoprotein with genetically separable functions in promoting viral DNA replication and the induction of oncogenic transformation. LT initiates replication by specifically binding to the viral origin and recruiting cellular replication factors. LT must also prepare the cell for replication by stimulating cell-cycle progression from G1 (or G0) into S-phase. It is this latter function of LT: the breakdown of cellular control mechanisms associated with LT induced cell-cycle progression that are likely to be main contributors to oncogenic transformation.

Transformation elicited by SV40, or its early gene products, can be assayed in multiple ways. Viral infection leads to transformation of a range of cultured rodent and human cells and induces tumors in newborn hamsters (Eddy *et al.*, 1962; Manfredi and Prives, 1994; Todaro *et al.*, 1966). Transfection of origin-deficient SV40 genomic DNA significantly enhances *in vitro* transformation of human cells, suggesting that viral replication presents an obstacle to stable transformation (Small *et al.*, 1982). The expression of LT alone, or in combination with ST, will oncogenically transform most rodent cell types. Studies have shown that expression of LT often allows cells to become immortalized as well as grow in reduced serum and until higher saturation density. Furthermore, these transformed cells escape contact inhibition as shown in focus formation assays. LT alone, or more frequently in conjunction with ST, induces anchorage-independent growth in soft agar and tumors in nude mice. Previously, several excellent reviews on LT have been published, some general in scope (Ahuja *et al.*, 2005; Cheng *et al.*, 2009; Fanning and Knippers, 1992; Pipas, 1992, 2009), while others emphasized transformation in different cell types (Manfredi and Prives, 1994), function of the DnaJ domain (Sullivan and Pipas, 2002), or transgenic models for tumorigenesis (Saenz Robles and Pipas, 2009).

Structure determination has yielded substantial insight to the different modules that together make up LT. The structure is known for the DnaJ domain and adjacent LxCxE motif (Kim *et al.*, 2001), the origin-binding domain (Luo *et al.*, 1996; Meinke *et al.*, 2007), the Zn binding domain and the ATPase/helicase domain (Li *et al.*, 2003). While it is believed that the independent domains are interconnected via flexible linkers, their relative positioning during replication is not known in detail. The structure of the host range domain at the extreme C-terminus is also not known. Insights to the structure has unraveled much greater detail about the functioning of the LT multicomponent machine, in part by revealing important contacts with known interactors like p53, and predicting binding surfaces for novel interactors (Ahuja *et al.*, 2009). Many activities of LT can be ascribed to discrete, linear regions of the protein. This is important for the mutational analysis of LT, which has yielded remarkable insight to its function in basic biological processes. We will traverse the LT molecule from the N- to the C-terminus to discuss each functional entity or binding protein (see Fig. 5 for a graphic depiction).

1. DnaJ DOMAIN

The demonstration that LT has recruited chaperone power to its repertoire stands as a landmark achievement (Campbell *et al.*, 1997; Cheetham *et al.*, 1992; Kelley and Landry, 1994; Srinivasan *et al.*, 1997). Previous data had strongly implicated the common region between LT and ST (referred to as T/t common domain) in transformation, but the mechanism was unknown (Marsilio *et al.*, 1991; Montano *et al.*, 1990; Peden and Pipas, 1992; Peden *et al.*, 1990; Pipas *et al.*, 1983; Zhu *et al.*, 1992). The LT N-terminus (first $\sim$70 amino acids) exhibits significant sequence homology with known DnaJ domains, including the canonical HPDK motif (Campbell *et al.*, 1997). Structural analysis of the LT N-terminus also indicates that it folds into a traditional DnaJ structure (Kim *et al.*, 2001). DnaJ proteins are molecular co-chaperones that recruit DnaK family chaperones (heat shock family) to perform functions such as protein folding, protein transport, or remodeling of protein complexes. The energy released from ATP hydrolysis by the DnaK family protein is used to act on a target protein. For LT it was demonstrated that the DnaJ domain binds the constitutively expressed Hsc70 and, as expected for a chaperone, significantly stimulates its ATPase activity (Campbell *et al.*, 1997; Srinivasan *et al.*, 1997; Sullivan and Pipas, 2002). Classical mutants in the DnaJ domain such as H42Q or D44N disrupt this binding. Functional assays demonstrated clearly that the DnaJ domain is critical for two distinct LT functions: to stimulate viral replication and to enhance oncogenic transformation via functional inactivation of the p107/p130 pRB family members (Campbell *et al.*, 1997; Srinivasan *et al.*, 1997; Stubdal *et al.*, 1996, 1997; Zalvide *et al.*, 1998). Strikingly, the LT DnaJ domain can be replaced by the cellular DnaJ protein HSJ1, and the chimeras

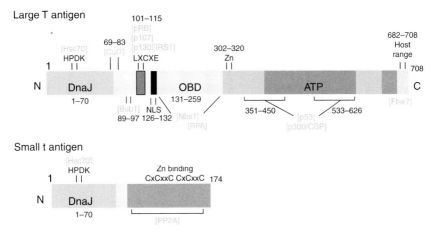

Fig. 5 *Schematic drawings of SV40 LT and ST. Landmark features of LT and ST.* At the N-terminus, the first ~70 amino acids contain the DnaJ domain with its canonical "HPDK" motif required for binding of Hsc70 chaperones. In LT, this is followed by binding sites for the Cul7 and Bub1 proteins, requiring amino acids 69–83 and 89–97, respectively. The LxCxE motif is critical for binding of pocket proteins pRB, p107, and p130, but also contributes to binding of IRS1. The NLS resides between amino acids 126 and 132. The OBD (origin-binding domain), besides conferring specific DNA binding activity to LT, also mediates binding of Nbs1 and replication protein A (RPA). The Zn^{2+}-binding element requires coordinating cysteine/histidine residues from amino acids 302 to 320 and allows LT to assemble into hexamers in an ATP-dependent manner. A bipartite binding site for p53 is present that requires amino acids 351–450 and 533–626; binding of p300/CBP transcriptional co-activators is bridged by p53. The ATPase domain labeled "ATP" is critical for LT helicase activity. The extreme C-terminus contains the host range domain as well as a binding site for the F-box protein Fbw7. The ST protein diverges from LT after the DnaJ domain, having instead at its C-terminus a binding site for the phosphatase PP2A. Cysteine clusters are responsible for coordinating Zn^{2+} and conformational stability.

retain LT DnaJ-dependent functions (Campbell *et al.*, 1997). However, there is a species-specific component, since the yeast Ydj1p or *E. coli* DnaJ proteins fail to functionally replace the LT DnaJ domain (Sullivan *et al.*, 2000b). This failure appeared to be because stimulation of Hsc70 ATPase was compromised, despite normal binding.

A contribution of the DnaJ domain to LT transformation is not observed in all assays for the transformed phenotype, but rather a subset. In assays for anchorage-independent growth, the D44N mutant shows no defect, and it also appears to immortalize mouse embryo fibroblasts normally (Stubdal *et al.*, 1997). However, to promote growth in low serum and to high saturation density, LT absolutely requires the DnaJ domain in cis with the pRB family binding site to disrupt complexes of p107/p130 with E2F4 (Stubdal *et al.*, 1997; Sullivan *et al.*, 2000a; Zalvide *et al.*, 1998).

The hyperphosphorylated forms of p107/p130 are also decreased via the DnaJ domain, and p130 is targeted for degradation (Stubdal et al., 1996, 1997). In human diploid fibroblasts, focus formation induced by LT + ST requires an intact DnaJ in LT, but not in ST, as shown using point mutants (Boyapati et al., 2003; Porras et al., 1996). A different study found no requirement for the LT DnaJ domain in human cell immortalization, transformation, or tumorigenesis; however, this was based on a combination of LT, ST, hTert, and oncogenic H-ras (Hahn et al., 2002).

In REF52 cell focus formation assays different DnaJ mutants exhibit variable defects. While the dl1135 mutant (deletion of residues 17–27) is completely defective, the D44N mutant only shows a modest defect in the context of full-length LT (Srinivasan et al., 1997). Strikingly, DnaJ point mutants like D44N exhibit much more pronounced defects when expressed in the context of LT1–136 (Beachy et al., 2002; Gjoerup et al., 2000; Srinivasan et al., 1997). The reason for the modest defects of DnaJ point mutants in a full-length background could be because D44N still retains partial DnaJ function (unlikely given data in the literature), because the deletion mutant is conformationally unfolded leading to loss of other LT functions, or because another transforming function exists within the LT N-terminus. Based on sequence homology it has been suggested that LT has a CR1 (conserved region 1 in adenovirus E1A) like sequence, which would be lost in dl1135 (Yaciuk et al., 1991). In E1A, the CR1 is important for transformation, because it is required for pRB and p300 binding (Berk, 2005). However, there is no evidence that the CR1-like sequence in LT functions in a similar manner. In fact, based on LT structural analysis, it is likely that the CR1-like sequence is buried within a hydrophobic core (Sullivan and Pipas, 2002).

While all the polyomavirus T antigens appear to have a functional DnaJ domain based on significant sequence homology, this is a unique feature that other viral oncoproteins like adenovirus E1A/E1B or human papillomavirus E6/E7 have not acquired. This is rather striking given the convergence of viral oncoproteins on targets like p53, p300/CBP, and the pRB family.

2. Cul7

Initial reports indicated that LT interacts with a 185 kDa cellular protein (Kohrman and Imperiale, 1992). The protein was subsequently identified as Cul7 from large-scale immunoprecipitations of LT1–135 from mouse NIH3T3 cells coupled with mass spectrometry (Ali et al., 2004). Cul7 is a member of an SCF (skp1, cullin, F-box) type E3 ubiquitin ligase complex that targets cellular proteins for proteasomal degradation. Cul7 assembles into complexes with Skp1, Fbxw8, and the Rbx1 ring finger protein (Sarikas et al., 2008). Mutations in Cul7 have been identified in the 3-M syndrome characterized by severe growth retardation (Huber *et al.*, 2005). Subsequent genetic analysis mapped the binding site for Cul7 to LT residues 69–83 (Kasper *et al.*, 2005). A deletion mutant from 69 to 83 is defective for Cul7 binding but exhibits normal binding of Bub1, pRB, and p53. Importantly, this mutant has a defect in growth to high density and in soft agar (Kasper *et al.*, 2005). Strikingly, the defect of the dl69-83 mutant in transformation can be rescued in Cul7-deficient mouse embryo fibroblasts, suggesting that LT inactivates at least some of the Cul7 functions. The critical substrates targeted for Cul7-mediated degradation and that are relevant for LT transformation have not been identified, but IRS1, a critical player in the insulin signaling pathway, was reported as a candidate (Xu *et al.*, 2008). Moreover, recent evidence revealed that LT via Cul7 targets the Mre11–Rad50–Nbs1 complex for degradation in an ATM-dependent manner during an SV40 infection (Zhao *et al.*, 2008). Currently it is not known if the interaction between LT and Cul7 occurs in a wide variety of cell types across species, and it is unclear if LT only acts to inhibit Cul7 or also to redirect it toward other substrates.

3. Bub1

A yeast two-hybrid screen, using the first 136 amino acids of LT as bait, yielded an interacting clone corresponding to the C-terminal kinase domain (amino acids 600–1085) of the mitotic checkpoint kinase Bub1 (Cotsiki *et al.*, 2004). Co-immunoprecipitation analysis has verified that the interaction occurs *in vivo* in mouse and human cells. A deletion mutant of LT, dl89-97, was identified that fails to bind Bub1 (Cotsiki *et al.*, 2004) yet retains binding of pRB, p53, and Cul7 (Hein *et al.*, 2009). Point mutagenesis identified tryptophans W91, W94, and W95 within a conserved sequence motif "WExWW" to be important for efficient binding. Focus formation assays in rat-1 cells demonstrated a strong correlation between LT binding to Bub1 and transformation (Cotsiki *et al.*, 2004). However, Bub1 binding was not required for cellular immortalization. Bub1 has primarily been shown to participate in the spindle checkpoint, a cellular surveillance mechanism that monitors whether kinetochores are properly attached to spindle microtubules (Meraldi and Sorger, 2005; Perera *et al.*, 2007) and delays the metaphase to anaphase transition. Thus, Bub1 is critical for maintaining genomic integrity. Interestingly, sporadic Bub1 mutations have been found in colorectal cancer, and reduced expression of Bub1 in mouse models leads to increased spontaneous tumorigenesis concomitant with aneuploidy (Cahill *et al.*, 1998; Jeganathan *et al.*, 2007; Schliekelman *et al.*, 2009).

It was demonstrated that LT via Bub1 binding leads to a compromise of the spindle checkpoint (Cotsiki *et al.*, 2004), not a total loss, which is known to be lethal (Perera *et al.*, 2007). Furthermore, when the spindle checkpoint is activated with nocodazole, cells expressing LT, not the dl89-97 mutant, are able to bypass the checkpoint and accumulate with >4 N DNA content. Recent results implicate Bub1 binding in the ability of LT to induce tetraploidy in BJ/tert human diploid fibroblasts (Hein *et al.*, 2009). Subsequent experiments revealed that shRNA-mediated knockdown of Bub1 expression leads to p53-dependent premature senescence in BJ/tert (Gjoerup *et al.*, 2007). Interestingly, expression of 17k with a pRB binding site mutation (designated "K1") also drove cells into premature senescence in a Bub1 binding-dependent manner, mimicking the outcome of Bub1 knockdown. Later experiments demonstrated that 17k K1 induces a DNA damage response via Bub1 binding (Hein *et al.*, 2009), consistent with recent reports that senescence in general is tied to an activated DNA damage response (Bartkova *et al.*, 2006; Di Micco *et al.*, 2006). These experiments also integrated LT function with p53, since the dl89-97 mutant is partially defective in stabilizing p53, likely because of a defect in p53 Ser15 phosphorylation (Hein *et al.*, 2009). The mechanism whereby LT is able to mount a DNA damage response via Bub1 binding is not known. Since the LT-induced DNA damage response occurs in the absence of the viral origin, one may speculate that the damage is associated with a deregulation of cellular DNA replication. Further experiments are likely to reveal exactly how LT acts on Bub1. It is unlikely to be a similar scenario to p53 or pRB where complete inactivation occurs, since total loss of Bub1 is lethal, even in somatic cells, due to catastrophic mitosis (Perera *et al.*, 2007). LT might redirect the Bub1 kinase toward a different set of substrates, of which other LT bound proteins are prime candidates.

It has long been known that LT induces both structural and numerical chromosome instability, but the mechanism has been elusive (Chang *et al.*, 1997; Friedrich *et al.*, 1992; Levine *et al.*, 1991; Ray and Kraemer, 1993; Ray *et al.*, 1990, 1992; Stewart and Bacchetti, 1991; Woods *et al.*, 1994). The interaction of LT with Bub1 may explain some of these observations, but more studies are clearly needed as additional LT binding proteins are likely to contribute. Whether LT-induced genomic instability, associated with gains and loss of specific chromosomes, contributes to long-term malignant transformation is unclear but of significant interest. Two prior observations suggest that it might. Most temperature-sensitive mutants of LT revert to the nontransformed phenotype at the nonpermissive temperature, but so-called A-type transformants remain transformed (Rassoulzadegan *et al.*, 1978; Seif and Martin, 1979). Likewise, when LT was expressed transgenically using an inducible system, hyperplasia and polyploidy could be reversed after 4 months, but no longer after 7 months, even though LT was turned off in both cases (Ewald *et al.*, 1996).

4. pRB FAMILY

The pRB protein is inactivated either by somatic or hereditary mutations leading to the pediatric tumor of the eye called retinoblastoma. The tumor suppressor concept underlying retinoblastoma was founded on Knudsen's original two-hit hypothesis (Knudson, 1971). By now, we know that every tumor virus has targeted pRB, or components of the pRB pathway, for inactivation. Indeed, almost every tumor, including those of nonviral origin, contains mutations in pRB or somewhere else within the pathway (Burkhart and Sage, 2008). Shortly after the report describing the cloning and characterization of the retinoblastoma protein as a 110-kDa nuclear phosphoprotein (Lee et al., 1987), it was demonstrated to be a critical transformation target of adenovirus E1A (Whyte et al., 1988). This was the first demonstration of a viral oncoprotein targeting a known tumor suppressor for inactivation. Subsequently, it was demonstrated that LT also binds pRB, and this is also required for malignant transformation by LT in a wide variety of cell types and assay systems (DeCaprio et al., 1988; Kalderon and Smith, 1984; Manfredi and Prives, 1994). Mutational analysis of LT clearly showed that a transforming function resides within amino acids 106–114 (Chen and Paucha, 1990). Taken together, it was found that LT from SV40 and MPyV, E1A and E7 all bind pRB (DeCaprio et al., 1988; Dyson et al., 1989; Munger et al., 1989; Whyte et al., 1988), and this requires a conserved sequence motif LxCxE.

The pRB protein consists of a number of domains, of which the A and B domains that together form the "short pocket" are critical for tumor suppression (Burkhart and Sage, 2008). Mutations and deletions in human malignancies map to this region, and it constitutes the binding site for the LxCxE motif. The pRB protein functions mainly as a transcriptional repressor. Although as many as 110 cellular proteins have been reported to bind pRB (Morris and Dyson, 2001), the E2F family of transcription factors appears to be the most important target in tumor suppression and cell cycle control. E2F members normally heterodimerize with members of the DP family to facilitate DNA binding. pRB effects transcriptional repression in large part by recruiting chromatin remodeling factors. These include histone deacetylases, hBRM, BRG1, and SUV39H1. The binding sites for LxCxE-containing proteins and E2F are distinct, although they both map within the pRB pocket domain. Thus, pRB can be simultaneously bound to E2F and chromatin remodeling factors to exert transcriptional repression.

After pRB was discovered, it was soon realized that two other members of the family exist, namely p107 and p130, and that these share structural and functional similarities with pRB. The pRB family is also referred to as "pocket proteins" due to their conserved interaction module. Interestingly, although partially redundant with pRB, p107 and p130 also possess unique, more specialized properties. Thus, a central spacer domain between A and B in p107 and p130, but not in pRB, binds to cyclins A and E in complexes with CDK2. While pRB is mutated in many types of cancer, mutations in p107 and p130 have rarely been observed in human malignancies. Although all of the pRB family members can bind E2F, they each bind different family members. Thus, pRB preferentially binds the "activating" E2Fs (E2F1-3), whereas p107/p130 preferentially bind "repressing" E2Fs (E2F4/5). Activating versus repressing species of E2F refers to whether they preferentially activate or repress their target genes. The downstream targets of E2F are extremely diverse, including genes involved in DNA replication, mitosis, DNA repair, differentiation, development, and apoptosis. The initial emphasis was on E2F targets involved in DNA replication, since these are *bona fide* targets of viruses creating an S-phase-like environment for viral replication.

Pocket proteins are clearly key players in negative growth regulation. Notably, different complexes form in different stages of the cell cycle. For example, p130-E2F complexes are found in G0, E2F4-p107 and E2F4-pRB complexes in G1, and free E2F1, -2 and -3 in S-phase. The levels of p107/p130 as well as several E2F family members are also cell cycle regulated. The pRB family is clearly involved in control of the restriction point, which is the point of no return in G1, when cells are committed to progression into S-phase. Additionally, pocket proteins are required for the G1–S checkpoint triggered following DNA damage. Recent experiments with mouse knockouts have demonstrated that loss of all three pocket proteins causes a failure to arrest in G1 following serum starvation, contact inhibition or DNA damage and is associated with cellular immortalization (Dannenberg *et al.*, 2000; Sage *et al.*, 2000). pRB is normally inactivated by sequential cyclin-dependent kinase phosphorylation, which leads to derepression of E2F. LT only targets the hypo- or underphosphorylated form of pRB, causing a loss of G1–S checkpoint control.

Accumulated evidence indicates that LT by binding all three pRB family members via its LxCxE motif causes the disruption of most, but perhaps not all, of their repressive complexes with E2F family members (Sullivan *et al.*, 2000a, 2004; Zalvide and DeCaprio, 1995; Zalvide *et al.*, 1998). As a consequence, many E2F target genes are activated or derepressed. Significantly, data from p107/p130 knockout cells has established very clearly that they are equally important targets for LT transformation as pRB. The DnaJ domain is absolutely required in addition to the LxCxE for inactivation of p107/p130 (Stubdal *et al.*, 1997; Sullivan *et al.*, 2000a; Zalvide *et al.*, 1998). Taken together, these observations have served to reinforce the notion that p107 and p130 are important in preventing tumorigenesis just like pRB.

Finally, it is important to take into consideration that there are certain effects of LT that are LxCxE dependent, but not DnaJ dependent, for example, the ability of LT to override p53 growth suppression and induce some properties of transformed cells (Gjoerup et al., 2000; Tevethia et al., 1997b). It has been demonstrated that LT through the LxCxE motif can partially overcome pRB-mediated repression of a heterologous promoter, possibly by disrupting interaction of pRB with HDAC corepressors that also bind via an LxCxE (Gjoerup et al., 2000).

Recent studies on E1A suggest that perhaps our model of LT acting merely to disrupt pRB/E2F complexes is too simplistic (DeCaprio, 2009). Reports indicate that E1A can be found at a number of different cellular promoters using chromatin immunoprecipitation techniques (Ferrari et al., 2008; Horwitz et al., 2008). E1A, like LT, binds a large number of cellular proteins including the pRB family and p300/CBP, which are histone acetyltransferases and coactivators (Berk, 2005). The implications are that E1A can recruit p300/CBP to pRB bound and repressed promoters with E2F sites and thereby turn on transcription. The reverse is likely also occurring. E1A could be recruiting repressive pRB complexes to promoters via an interaction of p300 with transcription factors bound upstream, effectively silencing the promoter. In other words, E1A by acting as an assembly platform that binds both transcriptionally repressive and activating factors can epigenetically reprogram the expression pattern of cellular genes, effectively tailoring the cellular environment to support viral replication. We can speculate that LT also may be bound to promoters repressed by pRB family members, and it might by recruitment of p300/CBP via p53 also be able to switch repressed promoters to an active state (DeCaprio, 2009). Currently there is little data to support this scenario for LT, but it is very plausible. Finally, it remains unclear if there are significant E2F-independent contributions to tumor suppression by pRB (Sellers et al., 1998), and whether these are specifically targeted by LT.

5. IRS1

Initial experiments demonstrated that LT cannot transform insulin-like growth factor I receptor (IGF-IR)-deficient cells (DeAngelis et al., 2006; Sell et al., 1993). Subsequently, it was shown that insulin receptor substrate 1 (IRS1), a key downstream target of IGF-I, is bound by LT, and together LT and IRS1 can transform IGF-IR null cells (D'Ambrosio et al., 1995; Fei et al., 1995). This implicated LT in IGF-I/insulin signaling pathways. A mutant lacking the first 250 amino acids failed to bind IRS1. LT causes the translocation of IRS1 from the cytoplasm to the nucleus (Prisco et al., 2002). Recently it was further shown that the pRB binding mutant K1 (E107K) is also defective for binding IRS1 and consequently defective in Akt activation (Yu and Alwine, 2008). It is possible that LT activation of Akt, leading to protection against apoptosis, is the main mechanism whereby IRS1 binding contributes to LT transformation. A mutant has not been identified yet that uniquely affects IRS1 binding.

6. Nbs1

LT binds the Nbs1 protein, a component of the MRN (Mre11, Rad50, Nbs1) complex that functions in DNA repair and possibly as a sensor of DNA double strand breaks (Lee and Paull, 2005; Wu et al., 2004; Zhao et al., 2008). Nbs1 is mutated in the Nijmegen breakage syndrome associated with chromosomal instability and increased cancer susceptibility. LT binding to the Nbs1 protein is believed to allow unrestrained firing of the viral origin, unlike cellular origins that are licensed to fire only once during S phase (Wu et al., 2004). The binding of LT to Nbs1 requires the LT origin-binding domain. An internal deletion mutant from 147 to 259 failed to bind Nbs1, whereas a deletion from 147 to 201 retained binding. During an SV40 infection, the MRN complex was shown to be degraded by LT in a Cul7 binding-dependent manner (Zhao et al., 2008). The MRN complex is also targeted for degradation by adenovirus (Stracker et al., 2002). Once a LT point mutant is generated that selectively affects only Nbs1 binding, a role of the LT/Nbs1 interaction in oncogenic transformation or genomic instability can be investigated.

7. p53

The p53 protein was first discovered as a LT binding protein in 1979 in studies that would revolutionize the fields of tumor virology and cancer research (Lane and Crawford, 1979; Linzer and Levine, 1979). Subsequent investigations have revealed that most viruses inactivate p53. Adenovirus E1B 55k binds and transcriptionally represses p53 and together with E4orf6 targets it for degradation (Nevels et al., 1997; Querido et al., 1997; Sarnow et al., 1982; Yew and Berk, 1992). Human papillomavirus E6 targets p53 for proteasomal degradation (Scheffner et al., 1990; Werness et al., 1990). While initially perceived as an oncogene, p53 was later realized to be in fact a tumor suppressor protein (Baker et al., 1990; Donehower et al., 1992; Eliyahu et al., 1989; Finlay et al., 1989; Malkin et al., 1990). As a confounding factor, the p53 clones that were used in initial studies contained mutations that made them oncogenic (Levine and Oren, 2009). In fact, p53 was shown to be mutated or deleted in at least 50% of human cancers (Hollstein et al., 1994). Thus, it is likely the most frequently mutated gene in human cancer, especially since many other tumors have mutations affecting other components of the p53 pathway. The p53 protein has earned the nickname "Guardian of the Genome" (Lane, 1992), because it responds to many types of cellular stress (e.g., genotoxic or oncogenic stress) by imposing either cell-cycle arrest, apoptosis, or senescence mediated by transcriptional control of a wealth of target genes. The p21^{CIP1} protein is a critical downstream target that mediates cell-cycle arrest by inhibiting cyclin-dependent kinases, pRB phosphorylation, and cell-cycle progression (el-Deiry et al., 1993).

LT binds p53 within its core DNA binding domain leading to loss of target gene activation (Bargonetti et al., 1992; Jiang et al., 1993; Mietz et al., 1992; Segawa et al., 1993). The p53-related proteins p63 and p73 do not appear to be targeted, at least not by direct binding (Marin et al., 1998). The region of LT required for p53 binding is bipartite residing in residues 351–450 and 533–626 intertwined with the helicase domain (Kierstead and Tevethia, 1993). Detailed insight on specific residues of LT and p53 required for their interaction, as well as the mechanism of p53 inactivation, became apparent when the crystal structure of the LT helicase domain in complex with p53 was reported (Lilyestrom et al., 2006). It is generally believed that LT induces unscheduled DNA synthesis via interaction with pRB family members. This aberrant proliferation response triggers p53-dependent apoptosis or growth arrest, which must be thwarted by LT. The p53 binding site is critical for mouse embryo fibroblast immortalization and for extension of lifespan in human diploid fibroblasts (Kierstead and Tevethia, 1993; Lin and Simmons, 1991; Zhu et al., 1991). In transformation assays using established cell lines, p53 binding is often not strictly required (Manfredi and Prives, 1994). In fact, LT1–121 can transform in some cell systems such as C3H10T1/2, albeit at reduced frequency (Srinivasan et al., 1989, 1997; Zhu et al., 1992). Interestingly, p53-mediated cell-cycle arrest is overcome by LT in a mainly pRB binding-dependent manner (Gjoerup et al., 2000; Quartin et al., 1994). This can be rationalized, because p53 in large part signals via p21^{CIP1} to pRB in order to induce cell-cycle arrest, reflecting important crosstalk between these two major tumor suppressor pathways.

Complexities of the LT/p53 interaction have gradually emerged. The traditional view that LT simply blocks p53 sequence-specific DNA binding via its interaction has been challenged by observations that mouse p53 bound to LT could contact DNA but not activate transcription, and human p53 bound to LT could both contact DNA and activate transcription (Sheppard et al., 1999). These contrasting views have not been resolved. There is also evidence that LT1–136, independent of p53 binding, can still inhibit p53-dependent transcription, although less efficiently than wild-type LT (Rushton et al., 1997). Perhaps even more confounding is the observation that LT potently stabilizes p53 (Deppert and Haug, 1986; Oren et al., 1981; Reich et al., 1983). Why would LT stabilize a major tumor suppressor protein, while at the same time functionally inactivating it? It was proposed that LT hijacks p53, essentially converting it from a tumor suppressor into an oncogene. In this scenario, presence of LT is not equivalent to a p53 null phenotype but actually causes a gain of function in p53 (Deppert et al., 1989; Hermannstadter et al., 2009; Tiemann and Deppert, 1994a,b). Work with LT transgenic mice also concluded that the presence of wild-type p53 enhanced tumor formation, consistent with a p53 gain of function (Herzig et al., 1999). Recently, it was demonstrated that the LT-p53 complex can have growth stimulatory properties, proposed to occur via activation of transcription from the insulin-like growth factor (IGF-1) promoter (Bocchetta et al., 2008). One might hypothesize that LT in part stabilizes p53 to gain access to p300/CBP that in turn act on promoters or acetylate other LT bound proteins (Borger and DeCaprio, 2006). Interestingly, several of the naturally occurring p53 mutants have been shown to exhibit a gain of function (Brosh and Rotter, 2009; Dittmer et al., 1993). Thus, a study of the LT stabilized form of p53 might also reveal insight that is germane to the mechanisms of tumorigenesis in nonviral lesions. The mechanistic basis for p53 stabilization by LT has not been fully elucidated, but likely involves induction of a DNA damage response via Bub1 binding (Hein et al., 2009). Thus, it might be a consequence of LT's strategy to co-opt the DNA damage response to enhance viral replication (Dahl et al., 2005; Hein et al., 2009; Shi et al., 2005; Zhao et al., 2008).

8. p300/CBP AND p400

E1A was first shown to bind p300/CBP, and this was linked to transformation and induction of cellular DNA synthesis in quiescent cells (Egan et al., 1988; Howe et al., 1990; Wang et al., 1993). Later, a pRB binding-deficient LT was shown to complement a p300 binding-deficient mutant of E1A for transformation, whereas the LT dl1135 mutant failed to complement (Yaciuk et al., 1991). This suggested that LT and E1A might target p300/CBP in analogous ways. While the initial studies suggested the binding site is N-terminal requiring the CR1-like sequence of LT, it was convincingly demonstrated later that a carboxy-terminal segment of LT (251–708) sufficed to bind, although at reduced efficiency compared to wild-type LT (Eckner et al., 1996; Lill et al., 1997). Furthermore, somewhat surprisingly, binding is indirect through p53 (Lill et al., 1997). Thus, binding to p300/CBP is absent in p53-deficient cells, but can be restored when p53 is reconstituted (Borger and DeCaprio, 2006). Given what we now know, it remains difficult to explain why dl1135 failed to complement, unless the DnaJ domain has to act on p300/CBP in some way.

CBP and p300 are large scaffolds involved in transcriptional regulation that contribute to many biological processes and are considered potential tumor suppressors (Gayther et al., 2000; Goodman and Smolik, 2000; Iyer et al., 2004). They act as adaptors or co-activators, in part via their intrinsic histone acetyltransferase (HAT) activity. Other proteins than histones can be acetylated by p300/CBP. Binding of LT to CBP results in specific acetylation of p53 and LT on K697, of which the significance is unclear (Borger and DeCaprio, 2006; Poulin et al., 2004). Transcription factors that are co-activated by p300/CBP include p53 and E2F. In one model LT recruits the p300/CBP co-activators to promoters that are normally repressed by pRB, thus eliciting promoter activation. Nevertheless, it remains unclear exactly how LT acts on p300/CBP. The initial report suggested that p300/CBP transcriptional activity is inhibited by LT (Eckner et al., 1996), but a complex scenario now seems more likely. With the recent implications that LT binding to p53 causes a gain of function, the most likely candidates to mediate this are p300/CBP, although the details have not yet been worked out. Although LT binds p300/CBP using p53 as an adaptor, recent work based on "patch" surface mutants of LT indicates that LT also directly makes contact with p300/CBP, and this is required for oncogenic transformation (Ahuja et al., 2009). The critical targets modulated by LT binding to p300/CBP largely remain to be identified, but one of them appears to be induction of c-myc (Singhal et al., 2008).

As with p300, p400 was first identified as a component of E1A immunocomplexes (Barbeau et al., 1994). Subsequent experiments confirmed that LT also binds p400, and the C-terminal fragment of LT251–708 retains binding (Lill et al., 1997). The binding site for p400 has not been further mapped to attempt to separate it from p300 binding. p400 is a SWI2/SNF2-related chromatin remodeling factor and interacts with a c-myc binding protein called TRRAP (Fuchs et al., 2001). Binding of the complex p400/TRRAP is critical for E1A transformation, perhaps because it is required to induce c-myc expression (Tworkowski et al., 2008). The p400 protein is also implicated in the p53–p21 cellular senescence pathway and p53-dependent apoptosis (Chan et al., 2005; Samuelson et al., 2005). Its role in LT-mediated transformation, if any, has not been elucidated.

B. Small t Antigen

Mutant SV40 viruses that fail to produce ST are viable but grow somewhat more slowly and produce less virus (Cicala et al., 1994; Sleigh et al., 1978). In vivo, expression of ST stimulates viral DNA replication (Cicala et al., 1994). Significant effects of ST are seen in some transformation assays. While LT expression is sufficient to induce foci in many rodent transformation systems, anchorage-independent growth often requires ST as well (Bikel et al., 1986; Bouck et al., 1978; Jog et al., 1990; Mungre et al., 1994; Sleigh et al., 1978). In the human cell system, ST is required both for focus formation and growth in soft agar (Chang et al., 1985; de Ronde et al., 1989; Hahn et al., 1999; Porras et al., 1996; Yu et al., 2001).

The demonstration that human cell transformation can be accomplished by defined genetic elements brought ST back in the spotlight, because the first example comprised a combination of LT, ST, oncogenic H-ras and hTERT (Hahn et al., 1999). LT is not sufficient to immortalize human cells (Neufeld et al., 1987), but the combination of LT and hTERT bypasses both senescence and crisis arising from telomere shortening, effectively creating immortal cells (Counter et al., 1998; Zhu et al., 1999).

Due to the splicing arrangement, ST has the first 82 amino acids in common with LT followed by 92 unique ones. As expected, ST has a functional DnaJ domain capable of stimulating the ATPase activity of DnaK proteins (Srinivasan et al., 1997). However, unlike the LT DnaJ domain, its function or target has not been elucidated, neither has it been implicated in ST-mediated transformation (Boyapati et al., 2003). While LT is mainly a nuclear protein, ST is distributed both in the cytoplasm and nucleus. ST binds Zn via two C-terminal CxCxxC clusters, which confer conformational stability (Turk et al., 1993). The majority of ST activities can be attributed to binding of the serine–threonine protein phosphatase PP2A (Rundell and Parakati, 2001; Sablina and Hahn, 2008; Skoczylas et al., 2004). PP2A is a heterotrimeric enzyme composed of an A scaffold subunit, a B regulatory subunit, and a C catalytic subunit. It is really a family of phosphatases, since there are two different A subunits, two C subunits and at least 17 B subunits, divided into four classes, that can assemble together into >100 different holoenzyme complexes (Sablina and Hahn, 2008). ST binds the A subunit and primarily acts to displace the B subunit or prevent it from binding the AC core complex (Pallas et al., 1990; Yang et al., 1991). In most cases, this leads to an inhibition of enzyme activity. However, ST most likely targets a specific subset of PP2A complexes that are still poorly defined (Sablina and Hahn, 2008). Although the DnaJ domain is not strictly required for A subunit binding, recent structural analyses indicate it most likely enhances binding to the AC complex and contributes to inhibition of PP2A activity (Chen et al., 2007; Cho et al., 2007). There are also examples like histone H1 (Yang et al., 1991), androgen receptor (Yang et al., 2007) and 4E-BP1 (Yu et al., 2005), where ST delivers PP2A to the substrate, thus mediating dephosphorylation rather than inhibiting it. In the case of 4E-BP1 dephosphorylation, this effect on the mTOR pathway causes inhibition of cap-dependent translation, which is consistent with a reduction in overall cellular translation at late stages of SV40 infection (Yu et al., 2005).

Human cell transformation assays, as well as other transformation assays that require ST, strictly depend on its binding to PP2A as shown using point and truncation mutants of ST (Hahn et al., 2002; Mungre et al., 1994). The morphological aspects of transformation might be mediated by the ability of ST via PP2A to induce profound disruption of the actin cytoskeleton (Nunbhakdi-Craig et al., 2003; Sontag and Sontag, 2006). PP2A acts in a myriad of proliferative or apoptotic pathways within the cell that are regulated by different isoforms of PP2A. Interestingly, knockdown of the PP2A B56Y subunit mimicked the effects of ST in anchorage-independent growth and tumorigenesis, suggesting that it is a key target (Chen *et al.*, 2004). Importantly, mutations in PP2A subunits have been found in human cancers, suggesting that some of these have tumor suppressor function (Arroyo and Hahn, 2005; Sablina and Hahn, 2008).

Some of the many and diverse targets of PP2A potentially involved in ST-induced tumorigenesis have been identified. One of these is c-myc, which is stabilized by ST via inhibition of PP2A-mediated dephosphorylation at S62 (Yeh *et al.*, 2004). Another important player is the phosphatidylinositol 3-kinase (PI3K) pathway that normally leads to activation of Akt, a prosurvival kinase, and Rac, a small GTP binding protein known to regulate proliferation and cytoskeletal organization (Skoczylas *et al.*, 2004; Zhao *et al.*, 2003). In human mammary epithelial cell transformation assays, ST could be replaced by constitutively active forms of PI3K or by Akt + Rac (Zhao *et al.*, 2003). Interestingly, a dominant negative subunit of PI3K blocked ST-dependent transformation in this system, although it does not appear that ST directly activates PI3K itself. ST also, again via PP2A binding, activates a number of kinases such as MAPK (Sontag *et al.*, 1993), Akt (Rodriguez-Viciana *et al.*, 2006; Yuan *et al.*, 2002), and PKCζ (Sontag *et al.*, 1997), the latter in turn leads to NFκB activation. MPyV ST has been demonstrated to be either pro- or anti-apoptotic dependent on the presence or absence of growth factors and the relative phosphorylation of Akt at sites T308 and S473 (Andrabi *et al.*, 2007). Studies have revealed that more than 30 kinases are regulated by PP2A (Millward *et al.*, 1999). ST can also activate the CAK kinase, resulting in cdk2 T160 phosphorylation, and together with cyclin E, promote active cyclin E/cdk2 complexes and concomitant DNA synthesis in quiescent cells (Sotillo *et al.*, 2008).

These observations may in part explain the long known ability of ST to induce proliferation, either alone or in some cell systems like human diploid fibroblasts, together with LT (Cicala *et al.*, 1994; Howe *et al.*, 1998; Porras *et al.*, 1999; Sontag *et al.*, 1993). Another important contribution comes from ST's ability to activate cellular promoters associated with proliferation control (Loeken *et al.*, 1988; Skoczylas *et al.*, 2004). The cyclin D1 promoter is activated via an AP1 site probably as a consequence of MAPK stimulation (Watanabe *et al.*, 1996). Also, ST activates the cyclin A promoter via a variant E2F site in its promoter (Skoczylas *et al.*, 2005). The cyclin-dependent kinase inhibitor $p27^{kip1}$ is much reduced upon ST expression, probably via proteasomal degradation, although the exact mechanism is unknown (Porras *et al.*, 1999). Thus, ST in a concerted manner targets the cyclin-dependent kinases involved in G1–S phase transition. However, high-level ST expression via an adenovirus vector also induces arrest in

G2 and prophase accompanied by a failure to assemble the spindle and centrosomes (Gaillard et al., 2001). Finally, microarray analysis has identified many genes involved in proliferation, apoptosis, integrin signaling, and immune responses whose expression is altered by ST (Moreno et al., 2004). Many genes are altered in expression via ST interaction with PP2A, but several other alterations are in fact independent of ST binding to PP2A, thus demonstrating that other functions of ST exist, although at the moment other cellular targets have not been identified.

C. Transgenic Model Systems

The transgenic expression of the SV40 early region in a wide variety of different tissues and organs has provided a wealth of insight to malignant progression (Ahuja et al., 2005; Saenz Robles and Pipas, 2009). Dependent on the cell system, LT alone, or in conjunction with ST, can induce either no stimulation of proliferation, hyperplasia, dysplasia, carcinoma, or a full-blown metastatic phenotype. It underscores the importance of cell context for tumor development. One of the most thoroughly analyzed systems involves targeted expression of LT to the choroid plexus epithelium (CPE) of the brain using the LPV transcriptional control elements (Brinster et al., 1984; Symonds et al., 1993, 1994). The mice succumb to aggressive tumors within 1–2 months. Interestingly, when an N-terminal fragment of LT is expressed, encoding amino acids 1–121, identical tumors are observed but they grow more slowly (Chen et al., 1992; Saenz Robles et al., 1994). This is due to apoptosis, and it is p53-dependent, because in a p53-deficient background LT1–121 induces rapid tumor growth with minimal apoptosis (Symonds et al., 1994). Tumor induction by LT1–121 was absolutely dependent on the LxCxE motif, implying a critical contribution from pocket protein inactivation. The underlying notion here is that LT1–121 binds pRB family members and induces unscheduled DNA synthesis. This leads via the E2F1 transcriptional pathway to induction of a p53-dependent checkpoint poised to eliminate the tumor cells. It elegantly demonstrates the necessity in this system of inactivating both pRB and p53 pathways to induce proliferation while maintaining cell viability. Targeted expression of LT1–121 in the mammary gland demonstrated a very similar scenario to the CPE of p53-dependent apoptosis limiting tumor development (Simin et al., 2004). Interestingly, p53 inactivation by LT was also required for induction of B and T cell lymphomas, probably because p53-dependent apoptosis is prominent in these cell types (Saenz Robles et al., 1994; Symonds et al., 1993).

However, in a wide range of other cell types and tissues, p53 inactivation is not a requirement. Expression of LT1–121 in astrocytes leads to aberrant proliferation and extensive apoptosis that is not dependent on p53 but on functional PTEN (Xiao et al., 2002). Thus, inactivation of the pRB pathway by LT1–121 combined with PTEN loss accelerates astrocytomas. When expressed in the liver, LT1–121 is fully capable of inducing hepatocellular carcinomas (Bennoun et al., 1998). This is dependent on both pRB binding and an intact DnaJ domain. Similarly, another N-terminal construct LT1–127 induces pancreatic acinar carcinomas, with no requirement for p53 inactivation (Tevethia et al., 1997a). Expression of LT alone, or together with ST, via the probasin promoter has also been used in a prostate cancer model. When expressed in the prostate epithelium, hyperplasia, adenocarcinomas, and metastases are observed (Greenberg et al., 1995; Masumori et al., 2001). However, expression of LT1–121 induces prostate intraepithelial neoplasia and eventually adenocarcinoma (Hill et al., 2005). In this system, PTEN, rather than p53, inhibits tumor progression via apoptosis.

Expression of LT in intestinal enterocytes leads to hyperplasia, and with age this progresses to dysplasia (Hauft et al., 1992; Kim et al., 1993, 1994; Markovics et al., 2005). This is dependent on pRB family binding and a functional DnaJ domain, but p53 binding is dispensable since LT1–121 or LT1–136 behaves similarly to full-length LT, although they induce dysplasia with lower penetrance (Markovics et al., 2005; Rathi et al., 2007, 2009). In fact, there is no binding to or stabilization of p53 by LT in enterocytes, and a targeted deletion of p53 does not enhance hyperplasia (Markovics et al., 2005). Recent microarray experiments indicate that in enterocytes LT and LT1–136 regulate an almost identical set of genes via their DnaJ and pRB binding motifs (Rathi et al., 2009). The majority of these are likely targets of E2F, mainly E2F2 and E2F3a. This demonstrates that enterocyte proliferation and tumorigenesis is in large part regulated via the pRB/E2F pathway (Rathi et al., 2009).

Taken together, whereas pRB inactivation is always required for aberrant proliferation and various aspects of tumorigenesis in each LT transgenic model system, p53 inactivation plays a highly variable role in tumor development and progression. Transgenic model systems are powerful because they allow us to directly assess the individual contribution of different genes to tumorigenesis in different cell types. Furthermore, they often provide a model of human cancer in which therapeutic modalities might be tested. It is not entirely clear why LT expression across a range of different cell and tissue types elicits such different responses, but it may reflect the general heterogeneity of cancers. LT could, in addition to the well-characterized targets pRB and p53, interact with other unique binding proteins within each cell type that contribute to specific aspects of tumorigenesis (Saenz Robles and Pipas, 2009). Alternatively, LT interacts with the same set of proteins in each cell type, but the response is different, because these cellular proteins are wired differently dependent on the cell system. Although pRB inactivation is clearly important for tumorigenesis, it remains unclear if additional binding proteins within LT1–136 (Bub1, Cul7, IRS1?) might contribute to the malignant phenotype observed in the various systems.

V. CONCLUSION

In the past 50 years, polyomaviruses such as SV40 and MPyV have contributed in critical ways to our understanding of basic mechanisms and key cellular proteins involved in carcinogenesis (Atkin *et al.*, 2009). However, before the discovery of MCV, evidence to support the etiologic role of polyomaviruses in human cancers was tenuous. Since its discovery, accumulating data provides compelling evidence that MCV is causally associated with MCC. It is likely that future investigations into the biology of this new tumor virus will be generalizable to other settings, including gaining insights into the development of nonvirally induced cancers.

The recent surge in discovery of new polyomaviruses suggests that there are more to come. There is long-standing serologic evidence for another LPV-like polyomavirus residing in the human population, although it has not been identified (Brade *et al.*, 1981). Studies have suggested that 15–25% of the population is seropositive for an LPV-related virus (Kean *et al.*, 2009). For KIV and WUV, studies are very preliminary and have thus far focused only on detection of viral DNA and serologic seroprevalence. Future studies of KIV and WUV are likely to reveal if these viruses are indeed linked to human pathogenesis. The relationship between JCV and BKV and human cancers has been difficult to determine despite extensive investigations, yet *in vitro* assays as well as animal experiments have clearly established their potential for oncogenic transformation. The range of diseases associated with these human polyomaviruses show that they are highly significant human infections. The accumulated knowledge gained from the study of each individual virus synergizes to help understand each in their particular niche.

ACKNOWLEDGMENTS

We thank Brian Schaffhausen and Jim Pipas for their helpful suggestions and comments. O. Gjoerup is supported by NIH grant RO1 A1078926. Y. Chang is funded through NIH RO1 grants CA136363 and CA120726, and an American Cancer Society Research Professorship.

REFERENCES

Abend, J. R., Jiang, M., and Imperiale, M. J. (2009a). BK virus and human cancer: innocent until proven guilty. *Semin. Cancer Biol.* **19**, 252–260.

Abend, J. R., Joseph, A. E., Das, D., Campbell-Cecen, D. B., and Imperiale, M. J. (2009b). A truncated T antigen expressed from an alternatively spliced BK virus early mRNA. *J. Gen. Virol.* **90**, 1238–1245.

Agelli, M., and Clegg, L. X. (2003). Epidemiology of primary Merkel cell carcinoma in the United States. *J. Am. Acad. Dermatol.* **49**, 832–841.

Ahuja, D., Saenz-Robles, M. T., and Pipas, J. M. (2005). SV40 large T antigen targets multiple cellular pathways to elicit cellular transformation. *Oncogene* **24**, 7729–7745.

Ahuja, D., Rathi, A. V., Greer, A. E., Chen, X. S., and Pipas, J. M. (2009). A structure-guided mutational analysis of simian virus 40 large T antigen: identification of surface residues required for viral replication and transformation. *J. Virol.* **83**, 8781–8788.

Ali, S. H., Kasper, J. S., Arai, T., and DeCaprio, J. A. (2004). Cul7/p185/p193 binding to simian virus 40 large T antigen has a role in cellular transformation. *J. Virol.* **78**, 2749–2757.

Allander, T., Andreasson, K., Gupta, S., Bjerkner, A., Bogdanovic, G., Persson, M. A., Dalianis, T., Ramqvist, T., and Andersson, B. (2007). Identification of a third human polyomavirus. *J. Virol.* **81**, 4130–4136.

Andrabi, S., Gjoerup, O. V., Kean, J. A., Roberts, T. M., and Schaffhausen, B. (2007). Protein phosphatase 2A regulates life and death decisions via Akt in a context-dependent manner. *Proc. Natl. Acad. Sci. USA* **104**, 19011–19016.

Andres, C., Belloni, B., Puchta, U., Sander, C. A., and Flaig, M. J. (2009a). Prevalence of MCPyV in Merkel cell carcinoma and non-MCC tumors. *J. Cutan. Pathol.* **37**, 28–34.

Andres, C., Belloni, B., Puchta, U., Sander, C. A., and Flaig, M. J. (2009b). Re: Clinical factors associated with Merkel cell polyomavirus infection in Merkel cell carcinoma. *J. Natl. Cancer Inst.* **101**, 1655–1656 (Author reply 1656–7).

Arroyo, J. D., and Hahn, W. C. (2005). Involvement of PP2A in viral and cellular transformation. *Oncogene* **24**, 7746–7755.

Atkin, S. J., Griffin, B. E., and Dilworth, S. M. (2009). Polyoma virus and simian virus 40 as cancer models: history and perspectives. *Semin. Cancer Biol.* **19**, 211–217.

Babakir-Mina, M., Ciccozzi, M., Alteri, C., Polchi, P., Picardi, A., Greco, F., Lucarelli, G., Arcese, W., Perno, C. F., and Ciotti, M. (2009a). Excretion of the novel polyomaviruses KI and WU in the stool of patients with hematological disorders. *J. Med. Virol.* **81**, 1668–1673.

Babakir-Mina, M., Ciccozzi, M., Bonifacio, D., Bergallo, M., Costa, C., Cavallo, R., Di Bonito, L., Perno, C. F., and Ciotti, M. (2009b). Identification of the novel KI and WU polyomaviruses in human tonsils. *J. Clin. Virol.* **46**, 75–79.

Baker, S. J., Markowitz, S., Fearon, E. R., Willson, J. K., and Vogelstein, B. (1990). Suppression of human colorectal carcinoma cell growth by wild-type p53. *Science* **249**, 912–915.

Barbeau, D., Charbonneau, R., Whalen, S. G., Bayley, S. T., and Branton, P. E. (1994). Functional interactions within adenovirus E1A protein complexes. *Oncogene* **9**, 359–373.

Bargonetti, J., Reynisdottir, I., Friedman, P. N., and Prives, C. (1992). Site-specific binding of wild-type p53 to cellular DNA is inhibited by SV40 T antigen and mutant p53. *Genes Dev.* **6**, 1886–1898.

Bartkova, J., Rezaei, N., Liontos, M., Karakaidos, P., Kletsas, D., Issaeva, N., Vassiliou, L. V., Kolettas, E., Niforou, K., Zoumpourlis, V. C., Takaoka, M., Nakagawa, H., *et al.* (2006). Oncogene-induced senescence is part of the tumorigenesis barrier imposed by DNA damage checkpoints. *Nature* **444**, 633–637.

Beachy, T. M., Cole, S. L., Cavender, J. F., and Tevethia, M. J. (2002). Regions and activities of simian virus 40 T antigen that cooperate with an activated ras oncogene in transforming primary rat embryo fibroblasts. *J. Virol.* **76**, 3145–3157.

Becker, J. C., Houben, R., Ugurel, S., Trefzer, U., Pfohler, C., and Schrama, D. (2009). MC polyomavirus is frequently present in Merkel cell carcinoma of European patients. *J. Invest. Dermatol.* **129**, 248–250.

Behzad-Behbahani, A., Klapper, P. E., Vallely, P. J., Cleator, G. M., and Khoo, S. H. (2004). Detection of BK virus and JC virus DNA in urine samples from immunocompromised (HIV-infected) and immunocompetent (HIV-non-infected) patients using polymerase chain reaction and microplate hybridisation. *J. Clin. Virol.* **29**, 224–229.

Bennoun, M., Grimber, G., Couton, D., Seye, A., Molina, T., Briand, P., and Joulin, V. (1998). The amino-terminal region of SV40 large T antigen is sufficient to induce hepatic tumours in mice. *Oncogene* **17**, 1253–1259.

Berk, A. J. (2005). Recent lessons in gene expression, cell cycle control, and cell biology from adenovirus. *Oncogene* **24**, 7673–7685.

Bhatia, K., Modali, R., and Goedert, J. J. (2010). Merkel cell polyomavirus is not detected in mesotheliomas. *J. Clin. Virol.* **47**, 196–198.

Bialasiewicz, S., Whiley, D. M., Lambert, S. B., Jacob, K., Bletchly, C., Wang, D., Nissen, M. D., and Sloots, T. P. (2008). Presence of the newly discovered human polyomaviruses KI and WU in Australian patients with acute respiratory tract infection. *J. Clin. Virol.* **41**, 63–68.

Bialasiewicz, S., Lambert, S. B., Whiley, D. M., Nissen, M. D., and Sloots, T. P. (2009). Merkel cell polyomavirus DNA in respiratory specimens from children and adults. *Emerg. Infect. Dis.* **15**, 492–494.

Bikel, I., Mamon, H., Brown, E. L., Boltax, J., Agha, M., and Livingston, D. M. (1986). The t-unique coding domain is important to the transformation maintenance function of the simian virus 40 small t antigen. *Mol. Cell. Biol.* **6**, 1172–1178.

Bluemn, E. G., Paulson, K. G., Higgins, E. E., Sun, Y., Nghiem, P., and Nelson, P. S. (2009). Merkel cell polyomavirus is not detected in prostate cancers, surrounding stroma, or benign prostate controls. *J. Clin. Virol.* **44**, 164–166.

Bocchetta, M., Eliasz, S., De Marco, M. A., Rudzinski, J., Zhang, L., and Carbone, M. (2008). The SV40 large T antigen-p53 complexes bind and activate the insulin-like growth factor-I promoter stimulating cell growth. *Cancer Res.* **68**, 1022–1029.

Bofill-Mas, S., and Girones, R. (2003). Role of the environment in the transmission of JC virus. *J. Neurovirol.* **9**(Suppl. 1), 54–58.

Bofill-Mas, S., Pina, S., and Girones, R. (2000). Documenting the epidemiologic patterns of polyomaviruses in human populations by studying their presence in urban sewage. *Appl. Environ. Microbiol.* **66**, 238–245.

Bonvoisin, C., Weekers, L., Xhignesse, P., Grosch, S., Milicevic, M., and Krzesinski, J. M. (2008). Polyomavirus in renal transplantation: a hot problem. *Transplantation* **85**, S42–S48.

Borger, D. R., and DeCaprio, J. A. (2006). Targeting of p300/CREB binding protein coactivators by simian virus 40 is mediated through p53. *J. Virol.* **80**, 4292–4303.

Bouck, N., Beales, N., Shenk, T., Berg, P., and di Mayorca, G. (1978). New region of the simian virus 40 genome required for efficient viral transformation. *Proc. Natl. Acad. Sci. USA* **75**, 2473–2477.

Boyapati, A., Wilson, M., Yu, J., and Rundell, K. (2003). SV40 17KT antigen complements dnaj mutations in large T antigen to restore transformation of primary human fibroblasts. *Virology* **315**, 148–158.

Brade, L., Muller-Lantzsch, N., and zur Hausen, H. (1981). B-lymphotropic papovavirus and possibility of infections in humans. *J. Med. Virol.* **6**, 301–308.

Brinster, R. L., Chen, H. Y., Messing, A., van Dyke, T., Levine, A. J., and Palmiter, R. D. (1984). Transgenic mice harboring SV40 T-antigen genes develop characteristic brain tumors. *Cell* **37**, 367–379.

Brosh, R., and Rotter, V. (2009). When mutants gain new powers: news from the mutant p53 field. *Nat. Rev. Cancer* **9**, 701–713.

Burkhart, D. L., and Sage, J. (2008). Cellular mechanisms of tumour suppression by the retinoblastoma gene. *Nat. Rev. Cancer* **8**, 671–682.

Busam, K. J., Jungbluth, A. A., Rekthman, N., Coit, D., Pulitzer, M., Bini, J., Arora, R., Hanson, N. C., Tassello, J. A., Frosina, D., Moore, P., and Chang, Y. (2009). Merkel cell polyomavirus expression in Merkel cell carcinomas and its absence in combined tumors and pulmonary neuroendocrine carcinomas. *Am. J. Surg. Pathol.* **33**, 1378–1385.

Cahill, D. P., Lengauer, C., Yu, J., Riggins, G. J., Willson, J. K., Markowitz, S. D., Kinzler, K. W., and Vogelstein, B. (1998). Mutations of mitotic checkpoint genes in human cancers. *Nature* **392**, 300–303.

Campbell, K. S., Mullane, K. P., Aksoy, I. A., Stubdal, H., Zalvide, J., Pipas, J. M., Silver, P. A., Roberts, T. M., Schaffhausen, B. S., and DeCaprio, J. A. (1997). DnaJ/hsp40 chaperone domain of SV40 large T antigen promotes efficient viral DNA replication. *Genes Dev.* **11**, 1098–1110.

Carter, J. J., Paulson, K. G., Wipf, G. C., Miranda, D., Madeleine, M. M., Johnson, L. G., Lemos, B. D., Lee, S., Warcola, A. H., Iyer, J. G., Nghiem, P., and Galloway, D. A. (2009). Association of Merkel cell polyomavirus-specific antibodies with Merkel cell carcinoma. *J. Natl. Cancer Inst.* **101**, 1510–1522.

Chan, H. M., Narita, M., Lowe, S. W., and Livingston, D. M. (2005). The p400 E1A-associated protein is a novel component of the p53 → p21 senescence pathway. *Genes Dev.* **19**, 196–201.

Chang, L. S., Pan, S., Pater, M. M., and Di Mayorca, G. (1985). Differential requirement for SV40 early genes in immortalization and transformation of primary rat and human embryonic cells. *Virology* **146**, 246–261.

Chang, T. H., Ray, F. A., Thompson, D. A., and Schlegel, R. (1997). Disregulation of mitotic checkpoints and regulatory proteins following acute expression of SV40 large T antigen in diploid human cells. *Oncogene* **14**, 2383–2393.

Cheetham, M. E., Brion, J. P., and Anderton, B. H. (1992). Human homologues of the bacterial heat-shock protein DnaJ are preferentially expressed in neurons. *Biochem. J.* **284**(Pt 2), 469–476.

Chen, S., and Paucha, E. (1990). Identification of a region of simian virus 40 large T antigen required for cell transformation. *J. Virol.* **64**, 3350–3357.

Chen, J., Tobin, G. J., Pipas, J. M., and Van Dyke, T. (1992). T-antigen mutant activities in vivo: roles of p53 and pRB binding in tumorigenesis of the choroid plexus. *Oncogene* **7**, 1167–1175.

Chen, W., Possemato, R., Campbell, K. T., Plattner, C. A., Pallas, D. C., and Hahn, W. C. (2004). Identification of specific PP2A complexes involved in human cell transformation. *Cancer Cell* **5**, 127–136.

Chen, Y., Xu, Y., Bao, Q., Xing, Y., Li, Z., Lin, Z., Stock, J. B., Jeffrey, P. D., and Shi, Y. (2007). Structural and biochemical insights into the regulation of protein phosphatase 2A by small t antigen of SV40. *Nat. Struct. Mol. Biol.* **14**, 527–534.

Chen, Y., Bord, E., Tompkins, T., Miller, J., Tan, C. S., Kinkel, R. P., Stein, M. C., Viscidi, R. P., Ngo, L. H., and Koralnik, I. J. (2009). Asymptomatic reactivation of JC virus in patients treated with natalizumab. *N. Engl. J. Med.* **361**, 1067–1074.

Cheng, J., DeCaprio, J. A., Fluck, M. M., and Schaffhausen, B. S. (2009). Cellular transformation by Simian Virus 40 and Murine Polyoma Virus T antigens. *Semin. Cancer Biol.* **19**, 218–228.

Cho, U. S., Morrone, S., Sablina, A. A., Arroyo, J. D., Hahn, W. C., and Xu, W. (2007). Structural basis of PP2A inhibition by small t antigen. *PLoS Biol.* **5**, e202.

Cicala, C., Avantaggiati, M. L., Graessmann, A., Rundell, K., Levine, A. S., and Carbone, M. (1994). Simian virus 40 small-t antigen stimulates viral DNA replication in permissive monkey cells. *J. Virol.* **68**, 3138–3144.

Cimbaluk, D., Pitelka, L., Kluskens, L., and Gattuso, P. (2009). Update on human polyomavirus BK nephropathy. *Diagn. Cytopathol.* **37**, 773–779.

Cotsiki, M., Lock, R. L., Cheng, Y., Williams, G. L., Zhao, J., Perera, D., Freire, R., Entwistle, A., Golemis, E. A., Roberts, T. M., Jat, P. S., and Gjoerup, O. V. (2004). Simian virus 40 large T antigen targets the spindle assembly checkpoint protein Bub1. *Proc. Natl. Acad. Sci. USA* **101**, 947–952.

Counter, C. M., Hahn, W. C., Wei, W., Caddle, S. D., Beijersbergen, R. L., Lansdorp, P. M., Sedivy, J. M., and Weinberg, R. A. (1998). Dissociation among in vitro telomerase activity, telomere maintenance, and cellular immortalization. *Proc. Natl. Acad. Sci. USA* **95**, 14723–14728.

Dahl, J., You, J., and Benjamin, T. L. (2005). Induction and utilization of an ATM signaling pathway by polyomavirus. *J. Virol.* **79**, 13007–13017.

D'Ambrosio, C., Keller, S. R., Morrione, A., Lienhard, G. E., Baserga, R., and Surmacz, E. (1995). Transforming potential of the insulin receptor substrate 1. *Cell Growth Differ.* **6**, 557–562.

Daniels, R., Sadowicz, D., and Hebert, D. N. (2007). A very late viral protein triggers the lytic release of SV40. *PLoS Pathog.* **3**, e98.

Dannenberg, J. H., van Rossum, A., Schuijff, L., and te Riele, H. (2000). Ablation of the retinoblastoma gene family deregulates G(1) control causing immortalization and increased cell turnover under growth-restricting conditions. *Genes Dev.* **14**, 3051–3064.

DeAngelis, T., Chen, J., Wu, A., Prisco, M., and Baserga, R. (2006). Transformation by the simian virus 40 T antigen is regulated by IGF-I receptor and IRS-1 signaling. *Oncogene* **25**, 32–42.

DeCaprio, J. A. (2009). How the Rb tumor suppressor structure and function was revealed by the study of Adenovirus and SV40. *Virology* **384**, 274–284.

DeCaprio, J. A., Ludlow, J. W., Figge, J., Shew, J. Y., Huang, C. M., Lee, W. H., Marsilio, E., Paucha, E., and Livingston, D. M. (1988). SV40 large tumor antigen forms a specific complex with the product of the retinoblastoma susceptibility gene. *Cell* **54**, 275–283.

Deppert, W., and Haug, M. (1986). Evidence for free and metabolically stable p53 protein in nuclear subfractions of simian virus 40-transformed cells. *Mol. Cell. Biol.* **6**, 2233–2240.

Deppert, W., Steinmayer, T., and Richter, W. (1989). Cooperation of SV40 large T antigen and the cellular protein p53 in maintenance of cell transformation. *Oncogene* **4**, 1103–1110.

de Ronde, A., Sol, C. J., van Strien, A., ter Schegget, J., and van der Noordaa, J. (1989). The SV40 small t antigen is essential for the morphological transformation of human fibroblasts. *Virology* **171**, 260–263.

Di Micco, R., Fumagalli, M., Cicalese, A., Piccinin, S., Gasparini, P., Luise, C., Schurra, C., Garre, M., Nuciforo, P. G., Bensimon, A., Maestro, R., Pelicci, P. G., *et al.* (2006). Oncogene-induced senescence is a DNA damage response triggered by DNA hyper-replication. *Nature* **444**, 638–642.

Dittmer, D., Pati, S., Zambetti, G., Chu, S., Teresky, A. K., Moore, M., Finlay, C., and Levine, A. J. (1993). Gain of function mutations in p53. *Nat. Genet.* **4**, 42–46.

Donehower, L. A., Harvey, M., Slagle, B. L., McArthur, M. J., Montgomery, C. A., Jr., Butel, J. S., and Bradley, A. (1992). Mice deficient for p53 are developmentally normal but susceptible to spontaneous tumours. *Nature* **356**, 215–221.

Drews, K., Bashir, T., and Dorries, K. (2000). Quantification of human polyomavirus JC in brain tissue and cerebrospinal fluid of patients with progressive multifocal leukoencephalopathy by competitive PCR. *J. Virol. Methods* **84**, 23–36.

Duncavage, E. J., Le, B. M., Wang, D., and Pfeifer, J. D. (2009a). Merkel cell polyomavirus: a specific marker for Merkel cell carcinoma in histologically similar tumors. *Am. J. Surg. Pathol.* **33**, 1771–1777.

Duncavage, E. J., Zehnbauer, B. A., and Pfeifer, J. D. (2009b). Prevalence of Merkel cell polyomavirus in Merkel cell carcinoma. *Mod. Pathol.* **22**, 516–521.

Dyson, N., Howley, P. M., Munger, K., and Harlow, E. (1989). The human papilloma virus-16 E7 oncoprotein is able to bind to the retinoblastoma gene product. *Science* **243**, 934–937.

Eash, S., Manley, K., Gasparovic, M., Querbes, W., and Atwood, W. J. (2006). The human polyomaviruses. *Cell. Mol. Life Sci.* **63**, 865–876.

Eckner, R., Ludlow, J. W., Lill, N. L., Oldread, E., Arany, Z., Modjtahedi, N., DeCaprio, J. A., Livingston, D. M., and Morgan, J. A. (1996). Association of p300 and CBP with simian virus 40 large T antigen. *Mol. Cell. Biol.* **16**, 3454–3464.

Eddy, B. E., Borman, G. S., Grubbs, G. E., and Young, R. D. (1962). Identification of the oncogenic substance in rhesus monkey kidney cell culture as simian virus 40. *Virology* **17**, 65–75.

Egan, C., Jelsma, T. N., Howe, J. A., Bayley, S. T., Ferguson, B., and Branton, P. E. (1988). Mapping of cellular protein-binding sites on the products of early-region 1A of human adenovirus type 5. *Mol. Cell. Biol.* **8,** 3955–3959.

Egli, A., Infanti, L., Dumoulin, A., Buser, A., Samaridis, J., Stebler, C., Gosert, R., and Hirsch, H. H. (2009). Prevalence of polyomavirus BK and JC infection and replication in 400 healthy blood donors. *J. Infect. Dis.* **199,** 837–846.

el-Deiry, W. S., Tokino, T., Velculescu, V. E., Levy, D. B., Parsons, R., Trent, J. M., Lin, D., Mercer, W. E., Kinzler, K. W., and Vogelstein, B. (1993). WAF1, a potential mediator of p53 tumor suppression. *Cell* **75,** 817–825.

Eliyahu, D., Michalovitz, D., Eliyahu, S., Pinhasi-Kimhi, O., and Oren, M. (1989). Wild-type p53 can inhibit oncogene-mediated focus formation. *Proc. Natl. Acad. Sci. USA* **86,** 8763–8767.

Elphick, G. F., Querbes, W., Jordan, J. A., Gee, G. V., Eash, S., Manley, K., Dugan, A., Stanifer, M., Bhatnagar, A., Kroeze, W. K., Roth, B. L., and Atwood, W. J. (2004). The human polyomavirus, JCV, uses serotonin receptors to infect cells. *Science* **306,** 1380–1383.

Engels, E. A., Frisch, M., Goedert, J. J., Biggar, R. J., and Miller, R. W. (2002). Merkel cell carcinoma and HIV infection. *Lancet* **359,** 497–498.

Erickson, K. D., Garcea, R. L., and Tsai, B. (2009). Ganglioside GT1b is a putative host cell receptor for the Merkel cell polyomavirus. *J. Virol.* **83,** 10275–10279.

Ewald, D., Li, M., Efrat, S., Auer, G., Wall, R. J., Furth, P. A., and Hennighausen, L. (1996). Time-sensitive reversal of hyperplasia in transgenic mice expressing SV40 T antigen. *Science* **273,** 1384–1386.

Fanning, E., and Knippers, R. (1992). Structure and function of simian virus 40 large tumor antigen. *Annu. Rev. Biochem.* **61,** 55–85.

Fei, Z. L., D'Ambrosio, C., Li, S., Surmacz, E., and Baserga, R. (1995). Association of insulin receptor substrate 1 with simian virus 40 large T antigen. *Mol. Cell. Biol.* **15,** 4232–4239.

Feng, H., Taylor, J. L., Benos, P. V., Newton, R., Waddell, K., Lucas, S. B., Chang, Y., and Moore, P. S. (2007). Human transcriptome subtraction by using short sequence tags to search for tumor viruses in conjunctival carcinoma. *J. Virol.* **81,** 11332–11340.

Feng, H., Shuda, M., Chang, Y., and Moore, P. S. (2008). Clonal integration of a polyomavirus in human Merkel cell carcinoma. *Science* **319,** 1096–1100.

Ferrari, R., Pellegrini, M., Horwitz, G. A., Xie, W., Berk, A. J., and Kurdistani, S. K. (2008). Epigenetic reprogramming by adenovirus e1a. *Science* **321,** 1086–1088.

Finlay, C. A., Hinds, P. W., and Levine, A. J. (1989). The p53 proto-oncogene can act as a suppressor of transformation. *Cell* **57,** 1083–1093.

Foulongne, V., Dereure, O., Kluger, N., Moles, J. P., Guillot, B., and Segondy, M. (2009). Merkel cell polyomavirus DNA detection in lesional and nonlesional skin from patients with Merkel cell carcinoma or other skin diseases. *Br. J. Dermatol.* **162,** 59–63.

Friedrich, T. D., Laffin, J., and Lehman, J. M. (1992). Simian virus 40 large T-antigen function is required for induction of tetraploid DNA content during lytic infection. *J. Virol.* **66,** 4576–4579.

Fuchs, M., Gerber, J., Drapkin, R., Sif, S., Ikura, T., Ogryzko, V., Lane, W. S., Nakatani, Y., and Livingston, D. M. (2001). The p400 complex is an essential E1A transformation target. *Cell* **106,** 297–307.

Gaillard, S., Fahrbach, K. M., Parkati, R., and Rundell, K. (2001). Overexpression of simian virus 40 small-T antigen blocks centrosome function and mitotic progression in human fibroblasts. *J. Virol.* **75,** 9799–9807.

Gardner, S. D., Field, A. M., Coleman, D. V., and Hulme, B. (1971). New human papovavirus (B.K.) isolated from urine after renal transplantation. *Lancet* **1,** 1253–1257.

Garneski, K. M., Warcola, A. H., Feng, Q., Kiviat, N. B., Leonard, J. H., and Nghiem, P. (2009). Merkel cell polyomavirus is more frequently present in North American than Australian Merkel cell carcinoma tumors. *J. Invest. Dermatol.* **129,** 246–248.

Gaynor, A. M., Nissen, M. D., Whiley, D. M., Mackay, I. M., Lambert, S. B., Wu, G., Brennan, D. C., Storch, G. A., Sloots, T. P., and Wang, D. (2007). Identification of a novel polyomavirus from patients with acute respiratory tract infections. *PLoS Pathog.* **3**, e64.

Gayther, S. A., Batley, S. J., Linger, L., Bannister, A., Thorpe, K., Chin, S. F., Daigo, Y., Russell, P., Wilson, A., Sowter, H. M., Delhanty, J. D., Ponder, B. A., *et al.* (2000). Mutations truncating the EP300 acetylase in human cancers. *Nat. Genet.* **24**, 300–303.

Gjoerup, O., Chao, H., DeCaprio, J. A., and Roberts, T. M. (2000). pRB-dependent, J domain-independent function of simian virus 40 large T antigen in override of p53 growth suppression. *J. Virol.* **74**, 864–874.

Gjoerup, O. V., Wu, J., Chandler-Militello, D., Williams, G. L., Zhao, J., Schaffhausen, B., Jat, P. S., and Roberts, T. M. (2007). Surveillance mechanism linking Bub1 loss to the p53 pathway. *Proc. Natl. Acad. Sci. USA* **104**, 8334–8339.

Goh, S., Lindau, C., Tiveljung-Lindell, A., and Allander, T. (2009). Merkel cell polyomavirus in respiratory tract secretions. *Emerg. Infect. Dis.* **15**, 489–491.

Goodman, R. H., and Smolik, S. (2000). CBP/p300 in cell growth, transformation, and development. *Genes Dev.* **14**, 1553–1577.

Greenberg, N. M., DeMayo, F., Finegold, M. J., Medina, D., Tilley, W. D., Aspinall, J. O., Cunha, G. R., Donjacour, A. A., Matusik, R. J., and Rosen, J. M. (1995). Prostate cancer in a transgenic mouse. *Proc. Natl. Acad. Sci. USA* **92**, 3439–3443.

Gross, L. (1953). A filterable agent, recovered from Ak leukemic extracts, causing salivary gland carcinomas in C3H mice. *Proc. Soc. Exp. Biol. Med.* **83**, 414–421.

Hahn, W. C., Counter, C. M., Lundberg, A. S., Beijersbergen, R. L., Brooks, M. W., and Weinberg, R. A. (1999). Creation of human tumour cells with defined genetic elements. *Nature* **400**, 464–468.

Hahn, W. C., Dessain, S. K., Brooks, M. W., King, J. E., Elenbaas, B., Sabatini, D. M., DeCaprio, J. A., and Weinberg, R. A. (2002). Enumeration of the simian virus 40 early region elements necessary for human cell transformation. *Mol. Cell. Biol.* **22**, 2111–2123.

Hauft, S. M., Kim, S. H., Schmidt, G. H., Pease, S., Rees, S., Harris, S., Roth, K. A., Hansbrough, J. R., Cohn, S. M., Ahnen, D. J., *et al.* (1992). Expression of SV-40 T antigen in the small intestinal epithelium of transgenic mice results in proliferative changes in the crypt and reentry of villus-associated enterocytes into the cell cycle but has no apparent effect on cellular differentiation programs and does not cause neoplastic transformation. *J. Cell. Biol.* **117**, 825–839.

Hein, J., Boichuk, S., Wu, J., Cheng, Y., Freire, R., Jat, P. S., Roberts, T. M., and Gjoerup, O. V. (2009). Simian virus 40 large T antigen disrupts genome integrity and activates a DNA damage response via Bub1 binding. *J. Virol.* **83**, 117–127.

Hermannstadter, A., Ziegler, C., Kuhl, M., Deppert, W., and Tolstonog, G. V. (2009). Wild-type p53 enhances efficiency of Simian virus 40 large T-antigen induced cellular transformation. *J. Virol.* **83**, 10106–10118.

Herzig, M., Novatchkova, M., and Christofori, G. (1999). An unexpected role for p53 in augmenting SV40 large T antigen-mediated tumorigenesis. *Biol. Chem.* **380**, 203–211.

Hill, R., Song, Y., Cardiff, R. D., and Van Dyke, T. (2005). Heterogeneous tumor evolution initiated by loss of pRb function in a preclinical prostate cancer model. *Cancer Res.* **65**, 10243–10254.

Hodgson, N. C. (2005). Merkel cell carcinoma: changing incidence trends. *J. Surg. Oncol.* **89**, 1–4.

Hollstein, M., Rice, K., Greenblatt, M. S., Soussi, T., Fuchs, R., Sorlie, T., Hovig, E., Smith-Sorensen, B., Montesano, R., and Harris, C. C. (1994). Database of p53 gene somatic mutations in human tumors and cell lines. *Nucleic Acids Res.* **22**, 3551–3555.

Horwitz, G. A., Zhang, K., McBrian, M. A., Grunstein, M., Kurdistani, S. K., and Berk, A. J. (2008). Adenovirus small e1a alters global patterns of histone modification. *Science* **321**, 1084–1085.

Howe, J. A., Mymryk, J. S., Egan, C., Branton, P. E., and Bayley, S. T. (1990). Retinoblastoma growth suppressor and a 300-kDa protein appear to regulate cellular DNA synthesis. *Proc. Natl. Acad. Sci. USA* **87**, 5883–5887.

Howe, A. K., Gaillard, S., Bennett, J. S., and Rundell, K. (1998). Cell cycle progression in monkey cells expressing simian virus 40 small t antigen from adenovirus vectors. *J. Virol.* **72**, 9637–9644.

Huber, C., Dias-Santagata, D., Glaser, A., O'Sullivan, J., Brauner, R., Wu, K., Xu, X., Pearce, K., Wang, R., Uzielli, M. L., Dagoneau, N., Chemaitilly, W., et al. (2005). Identification of mutations in CUL7 in 3-M syndrome. *Nat. Genet.* **37**, 1119–1124.

Huebner, K., Santoli, D., Croce, C. M., and Koprowski, H. (1975). Characterization of defective SV40 isolated from SV40-transformed cells. *Virology* **63**, 512–522.

Iyer, N. G., Ozdag, H., and Caldas, C. (2004). p300/CBP and cancer. *Oncogene* **23**, 4225–4231.

Jeffers, L. K., Madden, V., and Webster-Cyriaque, J. (2009). BK virus has tropism for human salivary gland cells in vitro: implications for transmission. *Virology* **394**, 183–193.

Jeganathan, K., Malureanu, L., Baker, D. J., Abraham, S. C., and van Deursen, J. M. (2007). Bub1 mediates cell death in response to chromosome missegregation and acts to suppress spontaneous tumorigenesis. *J. Cell. Biol.* **179**, 255–267.

Jiang, D., Srinivasan, A., Lozano, G., and Robbins, P. D. (1993). SV40 T antigen abrogates p53-mediated transcriptional activity. *Oncogene* **8**, 2805–2812.

Jog, P., Joshi, B., Dhamankar, V., Imperiale, M. J., Rutila, J., and Rundell, K. (1990). Mutational analysis of simian virus 40 small-t antigen. *J. Virol.* **64**, 2895–2900.

Kalderon, D., and Smith, A. E. (1984). In vitro mutagenesis of a putative DNA binding domain of SV40 large-T. *Virology* **139**, 109–137.

Kantola, K., Sadeghi, M., Lahtinen, A., Koskenvuo, M., Aaltonen, L. M., Mottonen, M., Rahiala, J., Saarinen-Pihkala, U., Riikonen, P., Jartti, T., Ruuskanen, O., Soderlund-Venermo, M., et al. (2009). Merkel cell polyomavirus DNA in tumor-free tonsillar tissues and upper respiratory tract samples: implications for respiratory transmission and latency. *J. Clin. Virol.* **45**, 292–295.

Kasper, J. S., Kuwabara, H., Arai, T., Ali, S. H., and DeCaprio, J. A. (2005). Simian virus 40 large T antigen's association with the CUL7 SCF complex contributes to cellular transformation. *J. Virol.* **79**, 11685–11692.

Kassem, A., Schopflin, A., Diaz, C., Weyers, W., Stickeler, E., Werner, M., and Zur Hausen, A. (2008). Frequent detection of Merkel cell polyomavirus in human Merkel cell carcinomas and identification of a unique deletion in the VP1 gene. *Cancer Res.* **68**, 5009–5013.

Kassem, A., Technau, K., Kurz, A. K., Pantulu, D., Loning, M., Kayser, G., Stickeler, E., Weyers, W., Diaz, C., Werner, M., Nashan, D., and Zur Hausen, A. (2009). Merkel cell polyomavirus sequences are frequently detected in nonmelanoma skin cancer of immunosuppressed patients. *Int. J. Cancer.* **125**, 356–361.

Katano, H., Ito, H., Suzuki, Y., Nakamura, T., Sato, Y., Tsuji, T., Matsuo, K., Nakagawa, H., and Sata, T. (2009). Detection of Merkel cell polyomavirus in Merkel cell carcinoma and Kaposi's sarcoma. *J. Med. Virol.* **81**, 1951–1958.

Kean, J. M., Rao, S., Wang, M., and Garcea, R. L. (2009). Seroepidemiology of human polyomaviruses. *PLoS Pathog.* **5**, e1000363.

Kelley, W. L., and Landry, S. J. (1994). Chaperone power in a virus? *Trends Biochem. Sci.* **19**, 277–278.

Khalili, K., White, M. K., Sawa, H., Nagashima, K., and Safak, M. (2005). The agnoprotein of polyomaviruses: a multifunctional auxiliary protein. *J. Cell. Physiol.* **204**, 1–7.

Kierstead, T. D., and Tevethia, M. J. (1993). Association of p53 binding and immortalization of primary C57BL/6 mouse embryo fibroblasts by using simian virus 40 T-antigen mutants bearing internal overlapping deletion mutations. *J. Virol.* **67**, 1817–1829.

Kim, S. H., Roth, K. A., Moser, A. R., and Gordon, J. I. (1993). Transgenic mouse models that explore the multistep hypothesis of intestinal neoplasia. *J. Cell. Biol.* **123,** 877–893.

Kim, S. H., Roth, K. A., Coopersmith, C. M., Pipas, J. M., and Gordon, J. I. (1994). Expression of wild-type and mutant simian virus 40 large tumor antigens in villus-associated enterocytes of transgenic mice. *Proc. Natl. Acad. Sci. USA* **91,** 6914–6918.

Kim, H. Y., Ahn, B. Y., and Cho, Y. (2001). Structural basis for the inactivation of retinoblastoma tumor suppressor by SV40 large T antigen. *EMBO J.* **20,** 295–304.

Kleinschmidt-DeMasters, B. K., and Tyler, K. L. (2005). Progressive multifocal leukoencephalopathy complicating treatment with natalizumab and interferon beta-1a for multiple sclerosis. *N. Engl. J. Med.* **353,** 369–374.

Knowles, W. A., Pipkin, P., Andrews, N., Vyse, A., Minor, P., Brown, D. W., and Miller, E. (2003). Population-based study of antibody to the human polyomaviruses BKV and JCV and the simian polyomavirus SV40. *J. Med. Virol.* **71,** 115–123.

Knudson, A. G., Jr. (1971). Mutation and cancer: statistical study of retinoblastoma. *Proc. Natl. Acad. Sci. USA* **68,** 820–823.

Kohrman, D. C., and Imperiale, M. J. (1992). Simian virus 40 large T antigen stably complexes with a 185-kilodalton host protein. *J. Virol.* **66,** 1752–1760.

Krynska, B., Del Valle, L., Croul, S., Gordon, J., Katsetos, C. D., Carbone, M., Giordano, A., and Khalili, K. (1999). Detection of human neurotropic JC virus DNA sequence and expression of the viral oncogenic protein in pediatric medulloblastomas. *Proc. Natl. Acad. Sci. USA* **96,** 11519–11524.

Kwun, H. J., Guastafierro, A., Shuda, M., Meinke, G., Bohm, A., Moore, P. S., and Chang, Y. (2009). The minimum replication origin of Merkel cell polyomavirus has a unique large T-antigen loading architecture and requires small T-antigen expression for optimal replication. *J. Virol.* **83,** 12118–12128.

Lane, D. P. (1992). Cancer. p53, guardian of the genome. *Nature* **358,** 15–16.

Lane, D. P., and Crawford, L. V. (1979). T antigen is bound to a host protein in SV40-transformed cells. *Nature* **278,** 261–263.

Langer-Gould, A., Atlas, S. W., Green, A. J., Bollen, A. W., and Pelletier, D. (2005). Progressive multifocal leukoencephalopathy in a patient treated with natalizumab. *N. Engl. J. Med.* **353,** 375–381.

Lee, J. H., and Paull, T. T. (2005). ATM activation by DNA double-strand breaks through the Mre11-Rad50-Nbs1 complex. *Science* **308,** 551–554.

Lee, W. H., Bookstein, R., Hong, F., Young, L. J., Shew, J. Y., and Lee, E. Y. (1987). Human retinoblastoma susceptibility gene: cloning, identification, and sequence. *Science* **235,** 1394–1399.

Levine, A. J., and Oren, M. (2009). The first 30 years of p53: growing ever more complex. *Nat. Rev. Cancer* **9,** 749–758.

Levine, D. S., Sanchez, C. A., Rabinovitch, P. S., and Reid, B. J. (1991). Formation of the tetraploid intermediate is associated with the development of cells with more than four centrioles in the elastase-simian virus 40 tumor antigen transgenic mouse model of pancreatic cancer. *Proc. Natl. Acad. Sci. USA* **88,** 6427–6431.

Li, D., Zhao, R., Lilyestrom, W., Gai, D., Zhang, R., DeCaprio, J. A., Fanning, E., Jochimiak, A., Szakonyi, G., and Chen, X. S. (2003). Structure of the replicative helicase of the oncoprotein SV40 large tumour antigen. *Nature* **423,** 512–518.

Lill, N. L., Tevethia, M. J., Eckner, R., Livingston, D. M., and Modjtahedi, N. (1997). p300 family members associate with the carboxyl terminus of simian virus 40 large tumor antigen. *J. Virol.* **71,** 129–137.

Lilyestrom, W., Klein, M. G., Zhang, R., Joachimiak, A., and Chen, X. S. (2006). Crystal structure of SV40 large T-antigen bound to p53: interplay between a viral oncoprotein and a cellular tumor suppressor. *Genes Dev.* **20,** 2373–2382.

Lin, J. Y., and Simmons, D. T. (1991). The ability of large T antigen to complex with p53 is necessary for the increased life span and partial transformation of human cells by simian virus 40. *J. Virol.* **65,** 6447–6453.

Linzer, D. I., and Levine, A. J. (1979). Characterization of a 54K dalton cellular SV40 tumor antigen present in SV40-transformed cells and uninfected embryonal carcinoma cells. *Cell* **17,** 43–52.

Loeken, M., Bikel, I., Livingston, D. M., and Brady, J. (1988). Trans-activation of RNA polymerase II and III promoters by SV40 small t antigen. *Cell* **55,** 1171–1177.

Low, J. A., Magnuson, B., Tsai, B., and Imperiale, M. J. (2006). Identification of gangliosides GD1b and GT1b as receptors for BK virus. *J. Virol.* **80,** 1361–1366.

Loyo, M., Guerrero-Preston, R., Brait, M., Hoque, M., Chuang, A., Kim, M., Sharma, R., Liegeois, N., Koch, W., Califano, J., Westra, W., and Sidransky, D. (2009). Quantitative detection of Merkel cell virus in human tissues and possible mode of transmission. *Int. J. Cancer* [Epub ahead of print].

Luo, X., Sanford, D. G., Bullock, P. A., and Bachovchin, W. W. (1996). Solution structure of the origin DNA-binding domain of SV40 T-antigen. *Nat. Struct. Biol.* **3,** 1034–1039.

Maginnis, M. S., and Atwood, W. J. (2009). JC virus: an oncogenic virus in animals and humans? *Semin. Cancer Biol.* **19,** 261–269.

Major, E. O. (2010). Progressive Multifocal Leukoencephalopathy in Patients on Immunomodulatory Therapies. *Annu. Rev. Med.* **61,** 35–47.

Malkin, D., Li, F. P., Strong, L. C., Fraumeni, J. F., Jr., Nelson, C. E., Kim, D. H., Kassel, J., Gryka, M. A., Bischoff, F. Z., Tainsky, M. A., et al. (1990). Germ line p53 mutations in a familial syndrome of breast cancer, sarcomas, and other neoplasms. *Science* **250,** 1233–1238.

Manfredi, J. J., and Prives, C. (1994). The transforming activity of simian virus 40 large tumor antigen. *Biochim. Biophys. Acta* **1198,** 65–83.

Maricich, S. M., Wellnitz, S. A., Nelson, A. M., Lesniak, D. R., Gerling, G. J., Lumpkin, E. A., and Zoghbi, H. Y. (2009). Merkel cells are essential for light-touch responses. *Science* **324,** 1580–1582.

Marin, M. C., Jost, C. A., Irwin, M. S., DeCaprio, J. A., Caput, D., and Kaelin, W. G., Jr. (1998). Viral oncoproteins discriminate between p53 and the p53 homolog p73. *Mol. Cell. Biol.* **18,** 6316–6324.

Markovics, J. A., Carroll, P. A., Robles, M. T., Pope, H., Coopersmith, C. M., and Pipas, J. M. (2005). Intestinal dysplasia induced by simian virus 40 T antigen is independent of p53. *J. Virol.* **79,** 7492–7502.

Marsilio, E., Cheng, S. H., Schaffhausen, B., Paucha, E., and Livingston, D. M. (1991). The T/t common region of simian virus 40 large T antigen contains a distinct transformation-governing sequence. *J. Virol.* **65,** 5647–5652.

Masumori, N., Thomas, T. Z., Chaurand, P., Case, T., Paul, M., Kasper, S., Caprioli, R. M., Tsukamoto, T., Shappell, S. B., and Matusik, R. J. (2001). A probasin-large T antigen transgenic mouse line develops prostate adenocarcinoma and neuroendocrine carcinoma with metastatic potential. *Cancer Res.* **61,** 2239–2249.

McQuaig, S. M., Scott, T. M., Harwood, V. J., Farrah, S. R., and Lukasik, J. O. (2006). Detection of human-derived fecal pollution in environmental waters by use of a PCR-based human polyomavirus assay. *Appl. Environ. Microbiol.* **72,** 7567–7574.

Meinke, G., Phelan, P., Moine, S., Bochkareva, E., Bochkarev, A., Bullock, P. A., and Bohm, A. (2007). The crystal structure of the SV40 T-antigen origin binding domain in complex with DNA. *PLoS Biol.* **5,** e23.

Meraldi, P., and Sorger, P. K. (2005). A dual role for Bub1 in the spindle checkpoint and chromosome congression. *EMBO J.* **24,** 1621–1633.

Mertz, K. D., Junt, T., Schmid, M., Pfaltz, M., and Kempf, W. (2009). Inflammatory monocytes are a reservoir for Merkel cell polyomavirus. *J. Invest. Dermatol.* **130,** 1146–1151.

Mietz, J. A., Unger, T., Huibregtse, J. M., and Howley, P. M. (1992). The transcriptional transactivation function of wild-type p53 is inhibited by SV40 large T-antigen and by HPV-16 E6 oncoprotein. *EMBO J.* **11**, 5013–5020.

Millward, T. A., Zolnierowicz, S., and Hemmings, B. A. (1999). Regulation of protein kinase cascades by protein phosphatase 2A. *Trends Biochem. Sci.* **24**, 186–191.

Monaco, M. C., Jensen, P. N., Hou, J., Durham, L. C., and Major, E. O. (1998a). Detection of JC virus DNA in human tonsil tissue: evidence for site of initial viral infection. *J. Virol.* **72**, 9918–9923.

Monaco, M. C., Shin, J., and Major, E. O. (1998b). JC virus infection in cells from lymphoid tissue. *Dev. Biol. Stand.* **94**, 115–122.

Montano, X., Millikan, R., Milhaven, J. M., Newsom, D. A., Ludlow, J. W., Arthur, A. K., Fanning, E., Bikel, I., and Livingston, D. M. (1990). Simian virus 40 small tumor antigen and an amino-terminal domain of large tumor antigen share a common transforming function. *Proc. Natl. Acad. Sci. USA* **87**, 7448–7452.

Moreno, C. S., Ramachandran, S., Ashby, D. G., Laycock, N., Plattner, C. A., Chen, W., Hahn, W. C., and Pallas, D. C. (2004). Signaling and transcriptional changes critical for transformation of human cells by simian virus 40 small tumor antigen or protein phosphatase 2A B56gamma knockdown. *Cancer Res.* **64**, 6978–6988.

Morris, E. J., and Dyson, N. J. (2001). Retinoblastoma protein partners. *Adv. Cancer Res.* **82**, 1–54.

Munger, K., Werness, B. A., Dyson, N., Phelps, W. C., Harlow, E., and Howley, P. M. (1989). Complex formation of human papillomavirus E7 proteins with the retinoblastoma tumor suppressor gene product. *EMBO J.* **8**, 4099–4105.

Mungre, S., Enderle, K., Turk, B., Porras, A., Wu, Y. Q., Mumby, M. C., and Rundell, K. (1994). Mutations which affect the inhibition of protein phosphatase 2A by simian virus 40 small-t antigen in vitro decrease viral transformation. *J. Virol.* **68**, 1675–1681.

Neufeld, D. S., Ripley, S., Henderson, A., and Ozer, H. L. (1987). Immortalization of human fibroblasts transformed by origin-defective simian virus 40. *Mol. Cell. Biol.* **7**, 2794–2802.

Nevels, M., Rubenwolf, S., Spruss, T., Wolf, H., and Dobner, T. (1997). The adenovirus E4orf6 protein can promote E1A/E1B-induced focus formation by interfering with p53 tumor suppressor function. *Proc. Natl. Acad. Sci. USA* **94**, 1206–1211.

Ng, S. C., Mertz, J. E., Sanden-Will, S., and Bina, M. (1985). Simian virus 40 maturation in cells harboring mutants deleted in the agnogene. *J. Biol. Chem.* **260**, 1127–1132.

Norja, P., Ubillos, I., Templeton, K., and Simmonds, P. (2007). No evidence for an association between infections with WU and KI polyomaviruses and respiratory disease. *J. Clin. Virol.* **40**, 307–311.

Nunbhakdi-Craig, V., Craig, L., Machleidt, T., and Sontag, E. (2003). Simian virus 40 small tumor antigen induces deregulation of the actin cytoskeleton and tight junctions in kidney epithelial cells. *J. Virol.* **77**, 2807–2818.

O'Donnell, P. H., Swanson, K., Josephson, M. A., Artz, A. S., Parsad, S. D., Ramaprasad, C., Pursell, K., Rich, E., Stock, W., and van Besien, K. (2009). BK virus infection is associated with hematuria and renal impairment in recipients of allogeneic hematopoetic stem cell transplants. *Biol. Blood Marrow. Transplant.* **15**, 1038–1048. e1.

Oren, M., Maltzman, W., and Levine, A. J. (1981). Post-translational regulation of the 54K cellular tumor antigen in normal and transformed cells. *Mol. Cell. Biol.* **1**, 101–110.

Padgett, B. L., Walker, D. L., ZuRhein, G. M., Eckroade, R. J., and Dessel, B. H. (1971). Cultivation of papova-like virus from human brain with progressive multifocal leucoencephalopathy. *Lancet* **1**, 1257–1260.

Pallas, D. C., Shahrik, L. K., Martin, B. L., Jaspers, S., Miller, T. B., Brautigan, D. L., and Roberts, T. M. (1990). Polyoma small and middle T antigens and SV40 small t antigen form stable complexes with protein phosphatase 2A. *Cell* **60**, 167–176.

Pastrana, D. V., Tolstov, Y. L., Becker, J. C., Moore, P. S., Chang, Y., and Buck, C. B. (2009). Quantitation of human seroresponsiveness to Merkel cell polyomavirus. *PLoS Pathog.* **5**, e1000578.

Peden, K. W., and Pipas, J. M. (1992). Simian virus 40 mutants with amino-acid substitutions near the amino terminus of large T antigen. *Virus Genes* **6**, 107–118.

Peden, K. W., Spence, S. L., Tack, L. C., Cartwright, C. A., Srinivasan, A., and Pipas, J. M. (1990). A DNA replication-positive mutant of simian virus 40 that is defective for transformation and the production of infectious virions. *J. Virol.* **64**, 2912–2921.

Perera, D., Tilston, V., Hopwood, J. A., Barchi, M., Boot-Handford, R. P., and Taylor, S. S. (2007). Bub1 maintains centromeric cohesion by activation of the spindle checkpoint. *Dev. Cell* **13**, 566–579.

Pina-Oviedo, S., De Leon-Bojorge, B., Cuesta-Mejias, T., White, M. K., Ortiz-Hidalgo, C., Khalili, K., and Del Valle, L. (2006). Glioblastoma multiforme with small cell neuronal-like component: association with human neurotropic JC virus. *Acta Neuropathol.* **111**, 388–396.

Pipas, J. M. (1992). Common and unique features of T antigens encoded by the polyomavirus group. *J. Virol.* **66**, 3979–3985.

Pipas, J. M. (2009). SV40: Cell transformation and tumorigenesis. *Virology* **384**, 294–303.

Pipas, J. M., Peden, K. W., and Nathans, D. (1983). Mutational analysis of simian virus 40 T antigen: isolation and characterization of mutants with deletions in the T-antigen gene. *Mol. Cell. Biol.* **3**, 203–213.

Porras, A., Bennett, J., Howe, A., Tokos, K., Bouck, N., Henglein, B., Sathyamangalam, S., Thimmapaya, B., and Rundell, K. (1996). A novel simian virus 40 early-region domain mediates transactivation of the cyclin A promoter by small-t antigen and is required for transformation in small-t antigen-dependent assays. *J. Virol.* **70**, 6902–6908.

Porras, A., Gaillard, S., and Rundell, K. (1999). The simian virus 40 small-t and large-T antigens jointly regulate cell cycle reentry in human fibroblasts. *J. Virol.* **73**, 3102–3107.

Poulin, D. L., Kung, A. L., and DeCaprio, J. A. (2004). p53 targets simian virus 40 large T antigen for acetylation by CBP. *J. Virol.* **78**, 8245–8253.

Prisco, M., Santini, F., Baffa, R., Liu, M., Drakas, R., Wu, A., and Baserga, R. (2002). Nuclear translocation of insulin receptor substrate-1 by the simian virus 40 T antigen and the activated type 1 insulin-like growth factor receptor. *J. Biol. Chem.* **277**, 32078–32085.

Quartin, R. S., Cole, C. N., Pipas, J. M., and Levine, A. J. (1994). The amino-terminal functions of the simian virus 40 large T antigen are required to overcome wild-type p53-mediated growth arrest of cells. *J. Virol.* **68**, 1334–1341.

Querido, E., Marcellus, R. C., Lai, A., Charbonneau, R., Teodoro, J. G., Ketner, G., and Branton, P. E. (1997). Regulation of p53 levels by the E1B 55-kilodalton protein and E4orf6 in adenovirus-infected cells. *J. Virol.* **71**, 3788–3798.

Rassoulzadegan, M., Perbal, B., and Cuzin, F. (1978). Growth control in simian virus 40-transformed rat cells: temperature-independent expression of the transformed phenotype in tsA transformants derived by agar selection. *J. Virol.* **28**, 1–5.

Rathi, A. V., Saenz Robles, M. T., and Pipas, J. M. (2007). Enterocyte proliferation and intestinal hyperplasia induced by simian virus 40 T antigen require a functional J domain. *J. Virol.* **81**, 9481–9489.

Rathi, A. V., Saenz Robles, M. T., Cantalupo, P. G., Whitehead, R. H., and Pipas, J. M. (2009). Simian virus 40 T-antigen-mediated gene regulation in enterocytes is controlled primarily by the Rb-E2F pathway. *J. Virol.* **83**, 9521–9531.

Ray, F. A., and Kraemer, P. M. (1993). Iterative chromosome mutation and selection as a mechanism of complete transformation of human diploid fibroblasts by SV40 T antigen. *Carcinogenesis* **14**, 1511–1516.

Ray, F. A., Peabody, D. S., Cooper, J. L., Cram, L. S., and Kraemer, P. M. (1990). SV40 T antigen alone drives karyotype instability that precedes neoplastic transformation of human diploid fibroblasts. *J. Cell. Biochem.* **42**, 13–31.

Ray, F. A., Meyne, J., and Kraemer, P. M. (1992). SV40 T antigen induced chromosomal changes reflect a process that is both clastogenic and aneuploidogenic and is ongoing throughout neoplastic progression of human fibroblasts. *Mutat. Res.* **284,** 265–273.

Reich, N. C., Oren, M., and Levine, A. J. (1983). Two distinct mechanisms regulate the levels of a cellular tumor antigen, p53. *Mol. Cell. Biol.* **3,** 2143–2150.

Ren, L., Gonzalez, R., Xie, Z., Zhang, J., Liu, C., Li, J., Li, Y., Wang, Z., Kong, X., Yao, Y., Hu, Y., Qian, S., *et al.* (2008). WU and KI polyomavirus present in the respiratory tract of children, but not in immunocompetent adults. *J. Clin. Virol.* **43,** 330–333.

Rodriguez-Viciana, P., Collins, C., and Fried, M. (2006). Polyoma and SV40 proteins differentially regulate PP2A to activate distinct cellular signaling pathways involved in growth control. *Proc. Natl. Acad. Sci. USA* **103,** 19290–19295.

Rundell, K., and Parakati, R. (2001). The role of the SV40 ST antigen in cell growth promotion and transformation. *Semin. Cancer Biol.* **11,** 5–13.

Rushton, J. J., Jiang, D., Srinivasan, A., Pipas, J. M., and Robbins, P. D. (1997). Simian virus 40 T antigen can regulate p53-mediated transcription independent of binding p53. *J. Virol.* **71,** 5620–5623.

Sablina, A. A., and Hahn, W. C. (2008). SV40 small T antigen and PP2A phosphatase in cell transformation. *Cancer Metastasis Rev.* **27,** 137–146.

Saenz Robles, M. T., and Pipas, J. M. (2009). T antigen transgenic mouse models. *Semin. Cancer Biol.* **19,** 229–235.

Saenz Robles, M. T., Symonds, H., Chen, J., and Van Dyke, T. (1994). Induction versus progression of brain tumor development: differential functions for the pRB- and p53-targeting domains of simian virus 40 T antigen. *Mol. Cell. Biol.* **14,** 2686–2698.

Sage, J., Mulligan, G. J., Attardi, L. D., Miller, A., Chen, S., Williams, B., Theodorou, E., and Jacks, T. (2000). Targeted disruption of the three Rb-related genes leads to loss of G(1) control and immortalization. *Genes Dev.* **14,** 3037–3050.

Samuelson, A. V., Narita, M., Chan, H. M., Jin, J., de Stanchina, E., McCurrach, M. E., Fuchs, M., Livingston, D. M., and Lowe, S. W. (2005). p400 is required for E1A to promote apoptosis. *J. Biol. Chem.* **280,** 21915–21923.

Sarikas, A., Xu, X., Field, L. J., and Pan, Z. Q. (2008). The cullin7 E3 ubiquitin ligase: a novel player in growth control. *Cell Cycle* **7,** 3154–3161.

Sarnow, P., Ho, Y. S., Williams, J., and Levine, A. J. (1982). Adenovirus E1b-58kd tumor antigen and SV40 large tumor antigen are physically associated with the same 54 kd cellular protein in transformed cells. *Cell* **28,** 387–394.

Sastre-Garau, X., Peter, M., Avril, M. F., Laude, H., Couturier, J., Rozenberg, F., Almeida, A., Boitier, F., Carlotti, A., Couturaud, B., and Dupin, N. (2009). Merkel cell carcinoma of the skin: pathological and molecular evidence for a causative role of MCV in oncogenesis. *J. Pathol.* **218,** 48–56.

Scheffner, M., Werness, B. A., Huibregtse, J. M., Levine, A. J., and Howley, P. M. (1990). The E6 oncoprotein encoded by human papillomavirus types 16 and 18 promotes the degradation of p53. *Cell* **63,** 1129–1136.

Schliekelman, M., Cowley, D. O., O'Quinn, R., Oliver, T. G., Lu, L., Salmon, E. D., and Van Dyke, T. (2009). Impaired Bub1 function in vivo compromises tension-dependent checkpoint function leading to aneuploidy and tumorigenesis. *Cancer Res.* **69,** 45–54.

Segawa, K., Minowa, A., Sugasawa, K., Takano, T., and Hanaoka, F. (1993). Abrogation of p53-mediated transactivation by SV40 large T antigen. *Oncogene* **8,** 543–548.

Seif, R., and Martin, R. G. (1979). Growth state of the cell early after infection with simian virus 40 determines whether the maintenance of transformation will be A-gene dependent or independent. *J. Virol.* **31,** 350–359.

Sell, C., Rubini, M., Rubin, R., Liu, J. P., Efstratiadis, A., and Baserga, R. (1993). Simian virus 40 large tumor antigen is unable to transform mouse embryonic fibroblasts lacking type 1 insulin-like growth factor receptor. *Proc. Natl. Acad. Sci. USA* **90,** 11217–11221.

Sellers, W. R., Novitch, B. G., Miyake, S., Heith, A., Otterson, G. A., Kaye, F. J., Lassar, A. B., and Kaelin, W. G., Jr. (1998). Stable binding to E2F is not required for the retinoblastoma protein to activate transcription, promote differentiation, and suppress tumor cell growth. *Genes Dev.* **12,** 95–106.

Seo, G. J., Fink, L. H., O'Hara, B., Atwood, W. J., and Sullivan, C. S. (2008). Evolutionarily conserved function of a viral microRNA. *J. Virol.* **82,** 9823–9828.

Sharp, C. P., Norja, P., Anthony, I., Bell, J. E., and Simmonds, P. (2009). Reactivation and mutation of newly discovered WU, KI, and Merkel cell carcinoma polyomaviruses in immunosuppressed individuals. *J. Infect. Dis.* **199,** 398–404.

Sheppard, H. M., Corneillie, S. I., Espiritu, C., Gatti, A., and Liu, X. (1999). New insights into the mechanism of inhibition of p53 by simian virus 40 large T antigen. *Mol. Cell. Biol.* **19,** 2746–2753.

Shi, Y., Dodson, G. E., Shaikh, S., Rundell, K., and Tibbetts, R. S. (2005). Ataxia-telangiectasia-mutated (ATM) is a T-antigen kinase that controls SV40 viral replication in vivo. *J. Biol. Chem.* **280,** 40195–40200.

Shuda, M., Feng, H., Kwun, H. J., Rosen, S. T., Gjoerup, O., Moore, P. S., and Chang, Y. (2008). T antigen mutations are a human tumor-specific signature for Merkel cell polyomavirus. *Proc. Natl. Acad. Sci. USA* **105,** 16272–16277.

Shuda, M., Arora, R., Kwun, H. J., Feng, H., Sarid, R., Fernandez-Figueras, M. T., Tolstov, Y., Gjoerup, O., Mansukhani, M. M., Swerdlow, S. H., Chaudhary, P. M., Kirkwood, J. M., *et al.* (2009). Human Merkel cell polyomavirus infection I. MCV T antigen expression in Merkel cell carcinoma, lymphoid tissues and lymphoid tumors. *Int. J. Cancer* **125,** 1243–1249.

Sihto, H., Kukko, H., Koljonen, V., Sankila, R., Bohling, T., and Joensuu, H. (2009). Clinical factors associated with Merkel cell polyomavirus infection in Merkel cell carcinoma. *J. Natl. Cancer Inst.* **101,** 938–945.

Simin, K., Wu, H., Lu, L., Pinkel, D., Albertson, D., Cardiff, R. D., and Van Dyke, T. (2004). pRb inactivation in mammary cells reveals common mechanisms for tumor initiation and progression in divergent epithelia. *PLoS Biol.* **2,** E22.

Singhal, G., Kadeppagari, R. K., Sankar, N., and Thimmapaya, B. (2008). Simian virus 40 large T overcomes p300 repression of c-Myc. *Virology* **377,** 227–232.

Skoczylas, C., Fahrbach, K. M., and Rundell, K. (2004). Cellular targets of the SV40 small-t antigen in human cell transformation. *Cell Cycle* **3,** 606–610.

Skoczylas, C., Henglein, B., and Rundell, K. (2005). PP2A-dependent transactivation of the cyclin A promoter by SV40 ST is mediated by a cell cycle-regulated E2F site. *Virology* **332,** 596–601.

Sleigh, M. J., Topp, W. C., Hanich, R., and Sambrook, J. F. (1978). Mutants of SV40 with an altered small t protein are reduced in their ability to transform cells. *Cell* **14,** 79–88.

Small, M. B., Gluzman, Y., and Ozer, H. L. (1982). Enhanced transformation of human fibroblasts by origin-defective simian virus 40. *Nature* **296,** 671–672.

Sontag, J. M., and Sontag, E. (2006). Regulation of cell adhesion by PP2A and SV40 small tumor antigen: an important link to cell transformation. *Cell. Mol. Life Sci.* **63,** 2979–2991.

Sontag, E., Fedorov, S., Kamibayashi, C., Robbins, D., Cobb, M., and Mumby, M. (1993). The interaction of SV40 small tumor antigen with protein phosphatase 2A stimulates the map kinase pathway and induces cell proliferation. *Cell* **75,** 887–897.

Sontag, E., Sontag, J. M., and Garcia, A. (1997). Protein phosphatase 2A is a critical regulator of protein kinase C zeta signaling targeted by SV40 small t to promote cell growth and NF-kappaB activation. *EMBO J.* **16,** 5662–5671.

Sotillo, E., Garriga, J., Kurimchak, A., and Grana, X. (2008). Cyclin E and SV40 small T antigen cooperate to bypass quiescence and contribute to transformation by activating CDK2 in human fibroblasts. *J. Biol. Chem.* **283,** 11280–11292.

Srinivasan, A., Peden, K. W., and Pipas, J. M. (1989). The large tumor antigen of simian virus 40 encodes at least two distinct transforming functions. *J. Virol.* **63,** 5459–5463.

Srinivasan, A., McClellan, A. J., Vartikar, J., Marks, I., Cantalupo, P., Li, Y., Whyte, P., Rundell, K., Brodsky, J. L., and Pipas, J. M. (1997). The amino-terminal transforming region of simian virus 40 large T and small t antigens functions as a J domain. *Mol. Cell. Biol.* **17**, 4761–4773.
Stewart, N., and Bacchetti, S. (1991). Expression of SV40 large T antigen, but not small t antigen, is required for the induction of chromosomal aberrations in transformed human cells. *Virology* **180**, 49–57.
Stracker, T. H., Carson, C. T., and Weitzman, M. D. (2002). Adenovirus oncoproteins inactivate the Mre11-Rad50-NBS1 DNA repair complex. *Nature* **418**, 348–352.
Stubdal, H., Zalvide, J., and DeCaprio, J. A. (1996). Simian virus 40 large T antigen alters the phosphorylation state of the RB-related proteins p130 and p107. *J. Virol.* **70**, 2781–2788.
Stubdal, H., Zalvide, J., Campbell, K. S., Schweitzer, C., Roberts, T. M., and DeCaprio, J. A. (1997). Inactivation of pRB-related proteins p130 and p107 mediated by the J domain of simian virus 40 large T antigen. *Mol. Cell. Biol.* **17**, 4979–4990.
Sullivan, C. S., and Pipas, J. M. (2002). T antigens of simian virus 40: molecular chaperones for viral replication and tumorigenesis. *Microbiol. Mol. Biol. Rev.* **66**, 179–202.
Sullivan, C. S., Cantalupo, P., and Pipas, J. M. (2000a). The molecular chaperone activity of simian virus 40 large T antigen is required to disrupt Rb-E2F family complexes by an ATP-dependent mechanism. *Mol. Cell. Biol.* **20**, 6233–6243.
Sullivan, C. S., Tremblay, J. D., Fewell, S. W., Lewis, J. A., Brodsky, J. L., and Pipas, J. M. (2000b). Species-specific elements in the large T-antigen J domain are required for cellular transformation and DNA replication by simian virus 40. *Mol. Cell. Biol.* **20**, 5749–5757.
Sullivan, C. S., Baker, A. E., and Pipas, J. M. (2004). Simian virus 40 infection disrupts p130-E2F and p107-E2F complexes but does not perturb pRb-E2F complexes. *Virology* **320**, 218–228.
Sullivan, C. S., Grundhoff, A. T., Tevethia, S., Pipas, J. M., and Ganem, D. (2005). SV40-encoded microRNAs regulate viral gene expression and reduce susceptibility to cytotoxic T cells. *Nature* **435**, 682–686.
Sullivan, C. S., Sung, C. K., Pack, C. D., Grundhoff, A., Lukacher, A. E., Benjamin, T. L., and Ganem, D. (2009). Murine Polyomavirus encodes a microRNA that cleaves early RNA transcripts but is not essential for experimental infection. *Virology* **387**, 157–167.
Sweet, B. H., and Hilleman, M. R. (1960). The vacuolating virus, S.V. 40. *Proc. Soc. Exp. Biol. Med.* **105**, 420–427.
Symonds, H. S., McCarthy, S. A., Chen, J., Pipas, J. M., and Van Dyke, T. (1993). Use of transgenic mice reveals cell-specific transformation by a simian virus 40 T-antigen amino-terminal mutant. *Mol. Cell. Biol.* **13**, 3255–3265.
Symonds, H., Krall, L., Remington, L., Saenz-Robles, M., Lowe, S., Jacks, T., and Van Dyke, T. (1994). p53-dependent apoptosis suppresses tumor growth and progression in vivo. *Cell* **78**, 703–711.
Tan, C. S., Dezube, B. J., Bhargava, P., Autissier, P., Wuthrich, C., Miller, J., and Koralnik, I. J. (2009). Detection of JC virus DNA and proteins in the bone marrow of HIV-positive and HIV-negative patients: implications for viral latency and neurotropic transformation. *J. Infect. Dis.* **199**, 881–888.
Tevethia, M. J., Bonneau, R. H., Griffith, J. W., and Mylin, L. (1997a). A simian virus 40 large T-antigen segment containing amino acids 1 to 127 and expressed under the control of the rat elastase-1 promoter produces pancreatic acinar carcinomas in transgenic mice. *J. Virol.* **71**, 8157–8166.
Tevethia, M. J., Lacko, H. A., Kierstead, T. D., and Thompson, D. L. (1997b). Adding an Rb-binding site to an N-terminally truncated simian virus 40 T antigen restores growth to high cell density, and the T common region in trans provides anchorage-independent growth and rapid growth in low serum concentrations. *J. Virol.* **71**, 1888–1896.

Tiemann, F., and Deppert, W. (1994a). Immortalization of BALB/c mouse embryo fibroblasts alters SV40 large T-antigen interactions with the tumor suppressor p53 and results in a reduced SV40 transformation-efficiency. *Oncogene* 9, 1907–1915.

Tiemann, F., and Deppert, W. (1994b). Stabilization of the tumor suppressor p53 during cellular transformation by simian virus 40: influence of viral and cellular factors and biological consequences. *J. Virol.* 68, 2869–2878.

Todaro, G. J., Green, H., and Swift, M. R. (1966). Susceptibility of human diploid fibroblast strains to transformation by SV40 virus. *Science* 153, 1252–1254.

Tolstov, Y. L., Pastrana, D. V., Feng, H., Becker, J. C., Jenkins, F. J., Moschos, S., Chang, Y., Buck, C. B., and Moore, P. S. (2009). Human Merkel cell polyomavirus infection II. MCV is a common human infection that can be detected by conformational capsid epitope immunoassays. *Int. J. Cancer* 125, 1250–1256.

Touze, A., Gaitan, J., Maruani, A., Le Bidre, E., Doussinaud, A., Clavel, C., Durlach, A., Aubin, F., Guyetant, S., Lorette, G., and Coursaget, P. (2009). Merkel cell polyomavirus strains in patients with merkel cell carcinoma. *Emerg. Infect. Dis.* 15, 960–962.

Trowbridge, P. W., and Frisque, R. J. (1995). Identification of three new JC virus proteins generated by alternative splicing of the early viral mRNA. *J. Neurovirol.* 1, 195–206.

Turk, B., Porras, A., Mumby, M. C., and Rundell, K. (1993). Simian virus 40 small-t antigen binds two zinc ions. *J. Virol.* 67, 3671–3673.

Tworkowski, K. A., Chakraborty, A. A., Samuelson, A. V., Seger, Y. R., Narita, M., Hannon, G. J., Lowe, S. W., and Tansey, W. P. (2008). Adenovirus E1A targets p400 to induce the cellular oncoprotein Myc. *Proc. Natl. Acad. Sci. USA* 105, 6103–6108.

Vago, L., Cinque, P., Sala, E., Nebuloni, M., Caldarelli, R., Racca, S., Ferrante, P., Trabottoni, G., and Costanzi, G. (1996). JCV-DNA and BKV-DNA in the CNS tissue and CSF of AIDS patients and normal subjects. Study of 41 cases and review of the literature. *J. Acquir. Immune. Defic. Syndr. Hum. Retrovirol.* 12, 139–146.

Van Assche, G., Van Ranst, M., Sciot, R., Dubois, B., Vermeire, S., Noman, M., Verbeeck, J., Geboes, K., Robberecht, W., and Rutgeerts, P. (2005). Progressive multifocal leukoencephalopathy after natalizumab therapy for Crohn's disease. *N. Engl. J. Med.* 353, 362–368.

Varga, E., Kiss, M., Szabo, K., and Kemeny, L. (2009). Detection of Merkel cell polyomavirus DNA in Merkel cell carcinomas. *Br. J. Dermatol.* 161, 930–932.

Wang, H. G., Rikitake, Y., Carter, M. C., Yaciuk, P., Abraham, S. E., Zerler, B., and Moran, E. (1993). Identification of specific adenovirus E1A N-terminal residues critical to the binding of cellular proteins and to the control of cell growth. *J. Virol.* 67, 476–488.

Watanabe, G., Howe, A., Lee, R. J., Albanese, C., Shu, I. W., Karnezis, A. N., Zon, L., Kyriakis, J., Rundell, K., and Pestell, R. G. (1996). Induction of cyclin D1 by simian virus 40 small tumor antigen. *Proc. Natl. Acad. Sci. USA* 93, 12861–12866.

Werness, B. A., Levine, A. J., and Howley, P. M. (1990). Association of human papillomavirus types 16 and 18 E6 proteins with p53. *Science* 248, 76–79.

Wetzels, C. T., Hoefnagel, J. G., Bakkers, J. M., Dijkman, H. B., Blokx, W. A., and Melchers, W. J. (2009). Ultrastructural proof of polyomavirus in Merkel cell carcinoma tumour cells and its absence in small cell carcinoma of the lung. *PLoS ONE* 4, e4958.

White, M. K., Gordon, J., Reiss, K., Del Valle, L., Croul, S., Giordano, A., Darbinyan, A., and Khalili, K. (2005). Human polyomaviruses and brain tumors. *Brain Res. Brain Res. Rev.* 50, 69–85.

Whyte, P., Buchkovich, K. J., Horowitz, J. M., Friend, S. H., Raybuck, M., Weinberg, R. A., and Harlow, E. (1988). Association between an oncogene and an anti-oncogene: the adenovirus E1A proteins bind to the retinoblastoma gene product. *Nature* 334, 124–129.

Woods, C., LeFeuvre, C., Stewart, N., and Bacchetti, S. (1994). Induction of genomic instability in SV40 transformed human cells: sufficiency of the N-terminal 147 amino acids of large T antigen and role of pRB and p53. *Oncogene* 9, 2943–2950.

Wu, X., Avni, D., Chiba, T., Yan, F., Zhao, Q., Lin, Y., Heng, H., and Livingston, D. (2004). SV40 T antigen interacts with Nbs1 to disrupt DNA replication control. *Genes Dev.* **18**, 1305–1316.

Xiao, A., Wu, H., Pandolfi, P. P., Louis, D. N., and Van Dyke, T. (2002). Astrocyte inactivation of the pRb pathway predisposes mice to malignant astrocytoma development that is accelerated by PTEN mutation. *Cancer Cell* **1**, 157–168.

Xu, X., Sarikas, A., Dias-Santagata, D. C., Dolios, G., Lafontant, P. J., Tsai, S. C., Zhu, W., Nakajima, H., Nakajima, H. O., Field, L. J., Wang, R., and Pan, Z. Q. (2008). The CUL7 E3 ubiquitin ligase targets insulin receptor substrate 1 for ubiquitin-dependent degradation. *Mol. Cell* **30**, 403–414.

Yaciuk, P., Carter, M. C., Pipas, J. M., and Moran, E. (1991). Simian virus 40 large-T antigen expresses a biological activity complementary to the p300-associated transforming function of the adenovirus E1A gene products. *Mol. Cell. Biol.* **11**, 2116–2124.

Yang, S. I., Lickteig, R. L., Estes, R., Rundell, K., Walter, G., and Mumby, M. C. (1991). Control of protein phosphatase 2A by simian virus 40 small-t antigen. *Mol. Cell. Biol.* **11**, 1988–1995.

Yang, C. S., Xin, H. W., Kelley, J. B., Spencer, A., Brautigan, D. L., and Paschal, B. M. (2007). Ligand binding to the androgen receptor induces conformational changes that regulate phosphatase interactions. *Mol. Cell. Biol.* **27**, 3390–3404.

Yeh, E., Cunningham, M., Arnold, H., Chasse, D., Monteith, T., Ivaldi, G., Hahn, W. C., Stukenberg, P. T., Shenolikar, S., Uchida, T., Counter, C. M., Nevins, J. R., *et al.* (2004). A signalling pathway controlling c-Myc degradation that impacts oncogenic transformation of human cells. *Nat. Cell Biol.* **6**, 308–318.

Yew, P. R., and Berk, A. J. (1992). Inhibition of p53 transactivation required for transformation by adenovirus early 1B protein. *Nature* **357**, 82–85.

Yu, Y., and Alwine, J. C. (2008). Interaction between simian virus 40 large T antigen and insulin receptor substrate 1 is disrupted by the K1 mutation, resulting in the loss of large T antigen-mediated phosphorylation of Akt. *J. Virol.* **82**, 4521–4526.

Yu, J., Boyapati, A., and Rundell, K. (2001). Critical role for SV40 small-t antigen in human cell transformation. *Virology* **290**, 192–198.

Yu, Y., Kudchodkar, S. B., and Alwine, J. C. (2005). Effects of simian virus 40 large and small tumor antigens on mammalian target of rapamycin signaling: small tumor antigen mediates hypophosphorylation of eIF4E-binding protein 1 late in infection. *J. Virol.* **79**, 6882–6889.

Yuan, H., Veldman, T., Rundell, K., and Schlegel, R. (2002). Simian virus 40 small tumor antigen activates AKT and telomerase and induces anchorage-independent growth of human epithelial cells. *J. Virol.* **76**, 10685–10691.

Zalvide, J., and DeCaprio, J. A. (1995). Role of pRb-related proteins in simian virus 40 large-T-antigen-mediated transformation. *Mol. Cell. Biol.* **15**, 5800–5810.

Zalvide, J., Stubdal, H., and DeCaprio, J. A. (1998). The J domain of simian virus 40 large T antigen is required to functionally inactivate RB family proteins. *Mol. Cell. Biol.* **18**, 1408–1415.

Zerrahn, J., Knippschild, U., Winkler, T., and Deppert, W. (1993). Independent expression of the transforming amino-terminal domain of SV40 large I antigen from an alternatively spliced third SV40 early mRNA. *EMBO J.* **12**, 4739–4746.

Zhao, J. J., Gjoerup, O. V., Subramanian, R. R., Cheng, Y., Chen, W., Roberts, T. M., and Hahn, W. C. (2003). Human mammary epithelial cell transformation through the activation of phosphatidylinositol 3-kinase. *Cancer Cell* **3**, 483–495.

Zhao, X., Madden-Fuentes, R. J., Lou, B. X., Pipas, J. M., Gerhardt, J., Rigell, C. J., and Fanning, E. (2008). Ataxia telangiectasia-mutated damage-signaling kinase- and proteasome-dependent destruction of Mre11-Rad50-Nbs1 subunits in Simian virus 40-infected primate cells. *J. Virol.* **82**, 5316–5328.

Zhu, J. Y., Abate, M., Rice, P. W., and Cole, C. N. (1991). The ability of simian virus 40 large T antigen to immortalize primary mouse embryo fibroblasts cosegregates with its ability to bind to p53. *J. Virol.* **65,** 6872–6880.

Zhu, J., Rice, P. W., Gorsch, L., Abate, M., and Cole, C. N. (1992). Transformation of a continuous rat embryo fibroblast cell line requires three separate domains of simian virus 40 large T antigen. *J. Virol.* **66,** 2780–2791.

Zhu, J., Wang, H., Bishop, J. M., and Blackburn, E. H. (1999). Telomerase extends the lifespan of virus-transformed human cells without net telomere lengthening. *Proc. Natl. Acad. Sci. USA* **96,** 3723–3728.

The Tyrosine Phosphatase Shp2 in Development and Cancer

Katja S. Grossmann,*,†,1,2 Marta Rosário,*,1,3 Carmen Birchmeier,† and Walter Birchmeier,*,4

*Department of Cancer Research, Max-Delbrueck-Center for Molecular Medicine, Berlin, Germany
†Neuroscience Department, Max-Delbrueck-Center for Molecular Medicine, Berlin, Germany

I. Introduction
II. Shp2 Activation and Signaling
III. Lessons from Human and Murine Gain-of-Function Mutations of Shp2
   A. Shp2 Gain-of-Function Mutations and Noonan Syndrome
   B. Shp2 Gain-of-Function Mutations in Cancer
IV. Lessons from Loss-of-Function Mutations of Shp2
   A. Shp2 LOF Mutations in Invertebrates and in *Xenopus*
   B. Null Mutations of Shp2 in Mice
   C. LOF Mutations of Shp2 in Humans: LEOPARD Syndrome
V. Organ Specific Roles of Shp2: Lessons from Mutations in Mice
   A. Shp2 and Heart Development
   B. Shp2 in T-Cell Development
   C. Shp2 in the CNS
   D. Shp2 and Neural Crest-Derived Tissues
   E. Shp2 in Pancreas, the Liver, and Mammary Gland
VI. Conclusions and Perspectives
   References

Deregulation of signaling pathways, through mutation or other molecular changes, can ultimately result in disease. The tyrosine phosphatase Shp2 has emerged as a major regulator of receptor tyrosine kinase (RTK) and cytokine receptor signaling. In the last decade, germline mutations in the human PTPN11 gene, encoding Shp2, were linked to Noonan (NS) and LEOPARD syndromes, two multisymptomatic developmental disorders that are characterized by short stature, craniofacial defects, cardiac defects, and mental retardation. Somatic Shp2 mutations are also associated with several types of human malignancies, such as the most common juvenile leukemia, juvenile myelomonocytic leukemia (JMML). Whereas NS and JMML are caused by gain-of-function (GOF) mutations of Shp2, loss-of-function (LOF) mutations are thought to be associated with LEOPARD syndrome. Animal models that carry conditional LOF and GOF

---

[1] These authors contributed equally to this work.
[2] Present address: Salk Institute for Biological Studies, La Jolla, California, USA
[3] Present address: Charité-Universitätsmedizin Berlin, Charitéplatz, Berlin, Germany
[4] Corresponding author: wbirch@mdc-berlin.de

mutations have allowed a better understanding of the mechanism of Shp2 function in disease, and shed light on the role of Shp2 in signaling pathways that control decisive events during embryonic development or during cellular transformation/tumorigenesis.
© 2010 Elsevier Inc.

I. INTRODUCTION

Mammalian Shp2 was independently identified in the early 1990s by several groups and was named Syp, SH-PTP2, SH-PTP3, PTP1D, or PTP2C (Adachi et al., 1992; Ahmad et al., 1993; Feng et al., 1993; Freeman et al., 1992; Vogel et al., 1993). Around the same time, the *Drosophila* homologue, Corkscrew (Csw), was identified in a genetic screen (Perkins et al., 1992). The encoded protein was shown to be a nonreceptor tyrosine phosphatase that contained N-terminal src-homology 2 (SH2) domains. Genetic analyses have demonstrated the importance of this tyrosine phosphatase for development and disease. Indeed, in humans germline and somatic mutations in PTPN11, the gene coding for Shp2, are associated with the Noonan (NS) and LEOPARD syndromes, as well as with a number of malignancies, most prominently juvenile myelomonocytic leukemia (JMML). The recent generation of mice that carry analogous mutations has led to a better understanding of the molecular mechanism of Shp2 in disease (Araki et al., 2004; Chan et al., 2009; Krenz et al., 2008; Nakamura et al., 2007). On the other hand, the major developmental defects resulting from loss-of-function (LOF) mutations in Shp2 in model organisms have emphasized the important and evolutionary conserved function of Shp2 in receptor tyrosine and cytokine signaling. This review focuses on the knowledge gained by the analysis of Shp2 mutations in mice (Table I) and lower organisms.

II. Shp2 ACTIVATION AND SIGNALING

Shp2 contains two N-terminally located *src-h*omology 2 domains (N-SH2 and C-SH2), a central phosphotyrosine phosphatase (PTP) domain, and a C-terminal tail with tyrosyl phosphorylation sites and a proline-rich motif (Fig. 1; reviewed in Neel et al., 2003; Rosário and Birchmeier, 2003). Both the SH2 and PTP domains are required for Shp2 function (Milarski and Saltiel, 1994; Noguchi et al., 1994; Tang et al., 1995). The C-terminal tyrosine residues of Shp2 appear to be phosphorylated in response to the activation of a subset of receptors, such as the PDGFR, upon which they can recruit signaling molecules such as Grb2 (Araki et al., 2003). The role of the proline-rich motif is unknown. Biochemical and structural studies show that

Table I Gain- and Loss-of-Function Mutations of Shp2 in Mice

Tissue	Strategy	Cre line	Shp2	Phenotype	References
Ubiquitous	Ubiquitous expression of Shp2 D61G	—	GOF	Craniofacial defects, heart defects, myeloproliferative disorder	Araki et al. (2004)
Ubiquitous	Inducible but ubiquitous expression of Shp2 D61Y	Mx-1 cre	GOF	Myeloproliferative disorder	Chan et al. (2009)
Ubiquitous	Ubiquitous expression of Shp2 N308D	—	GOF	Craniofacial defects, heart defects, myeloproliferative disorder	Araki et al. (2009)
Myocardial cells	βMHC specific expression of Shp2 Q79R	—	GOF	Ventricular compaction defects, ventricular septal defects	Nakamura et al. (2007)
Endocardial cells	Inducible expression of Shp2 Q79R	Tie2-cre	GOF	Enlarged endocardial cushion	Krenz et al. (2008)
Endocardial cells	Inducible expression of Shp2 D61Y	Tie2-cre	GOF	Enlarged endocardial cushion, ventricular septal defects, double-outlet right ventricle, thin myocardium	Araki et al. (2009)
Neural crest cells	Inducible expression of Shp2 Q79R	Wnt1-cre	GOF	Craniofacial defects	Nakamura et al. (2009b)
Ubiquitous	Shp2 null mutation	—	LOF	Inner cell mass death, reduced number of trophoblast giant cells, failure to yield trophoblast stem cells	Yang et al. (2006)
Myocardial and skeletal muscle cells	Inducible Shp2 null mutation	αMHC-cre, Mck-cre	LOF	Dilated cardiomyopathy, insulin resistance, glucose intolerance	Kontaridis et al. (2008), Princen et al. (2009)
Skeletal muscle	Inducible Shp2 null mutation	Mck-cre	LOF	Reduced myofiber size and number	Fornaro et al. (2006)
Neural crest cells	Inducible Shp2 null mutation	Wnt1-cre	LOF	Defective Schwann cell development, reduced myelination; enteric nervous system defects; craniofacial defects; failure of outflow tract septation, abnormal great arteries	Grossmann et al. (2009), Nakamura et al. (2009a)
T-cells	Inducible Shp2 null mutation	Lck-cre	LOF	Defective T-cells maturation and proliferation	Nguyen et al. (2006)

(*continues*)

Table I (continued)

Tissue	Strategy	Cre line	Shp2	Phenotype	References
CNS	Inducible Shp2 null mutation	Nestin-cre	LOF	Impaired corticogenesis, decreased neurogenesis, increased gliogenesis, reduced and less foliated cerebellum	Ke et al. (2007), Hagihara et al. (2009)
Cerebral cortex and hypothalamus	Inducible Shp2 null mutation	CamKII-cre, Cre3	LOF	Early onset obesity and diabetes	Zhang et al. (2004), Krajewska et al. (2008)
Pancreas	Inducible Shp2 null mutation	Pdx-1 cre	LOF	Decreased insulin production by β-cells and glucose intolerance	Zhang et al. (2009a)
Liver	Inducible Shp2 null mutation	Alb-cre	LOF	Defects in liver regeneration after partial hepatectomy	Bard-Chapeau et al. (2005, 2006)
Mammary gland	Inducible Shp2 null mutation	MMTV-cre	LOF	Impaired morphogenesis of alveolar structures during lactation	Ke et al. (2006)

Shown are the gain- and loss-of-function mutations of Shp2 that have been generated in mice using different strategies and cre-lines. The affected tissue, the phenotype and the references are indicated.

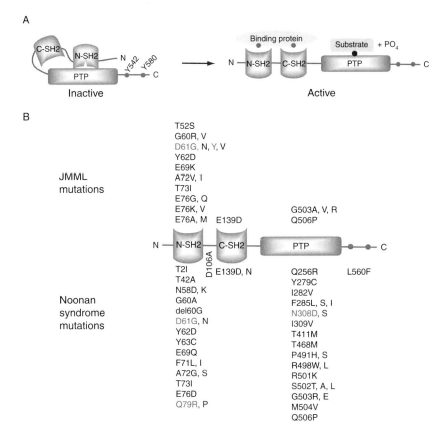

Fig. 1 Activation of Shp2 and mutations associated with Noonan syndrome and JMML. (A) Shp2 consists of two SH2 domains (N-SH2 and C-SH2), a central protein tyrosine phosphatase (PTP) domain, and two tyrosine residues in the C-terminus (Y542 and Y580). Shp2 is kept inactive in unstimulated cells through an interaction of the N-terminal SH2 domain with the phosphatase domain that blocks access to the catalytic site. Upon stimulation of cells, interaction of the SH2 domains with specific phosphotyrosine binding sites on activating proteins induces a conformational change in the N-SH2 domain that releases this autoinhibitory interaction and allows substrate access to the catalytic site. (B) Many mutations in the gene encoding for Shp2 have been described in Noonan syndrome and JMML patients. Most mutations reside in the N-terminal SH2 domain and the PTP domain. These mutations interfere with the inactive conformation of Shp2 and lead to an open and more readily activatable Shp2 protein. Mutations highlighted in magenta have been introduced into the mouse Shp2 gene. Adapted from Chan et al. (2008) and Neel et al. (2003).

in the resting state, Shp2 adopts an inactive "closed" configuration, in which a nonphosphotyrosine-dependent interaction between the N-terminal SH2 domain and the PTP domain precludes access of substrates to the catalytic site. Upon growth factor or cytokine stimulation, the interaction of the Shp2 SH2 domains with particular phosphorylated tyrosine residues on activating proteins induces a conformational change in the N-terminal SH2 domain, which opens the structure and relieves the autoinhibition (Barford and Neel, 1998; Hof et al., 1998). Shp2-activators include molecules such as the PDGF and gp130 receptors, but also a number of ubiquitous as well as tissue-specific adaptor proteins (Hadari et al., 1998; Maroun et al., 2000; Rosário et al., 2007; Schaeper et al., 2000).

The Ras/ERK MAP kinase pathway is a major pathway downstream of receptor tyrosine kinases (RTKs) and cytokine signaling that is modulated by Shp2 (reviewed in Feng, 2007; Neel et al., 2003; Rosário and Birchmeier, 2003). Specifically, Shp2 has been shown to promote the sustained activation of Ras and the downstream ERK MAP kinases. Several mechanisms have been identified by which Shp2 activates the Ras/ERK pathway (shown in Fig. 2). It should be noted that receptor and/or cell type specificity may determine how Shp2 regulates the Ras/ERK MAP kinase pathway.

Furthermore, numerous reports have linked Shp2 to the regulation of the JAK/Stat pathway (reviewed in Qu, 2002). In neural or hematopoietic cell types, Shp2 was reported to repress signaling either by dephosphorylation of

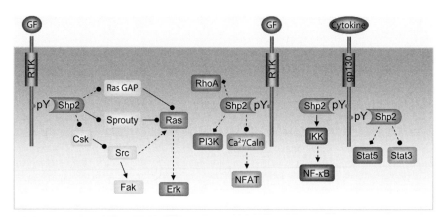

Fig. 2 Shp2-regulated signaling pathways. The major signaling pathway activated by Shp2 downstream of receptor tyrosine kinases (RTKs) and growth factors (GF) is the Ras/ERK MAP kinase cascade. Shp2 activates Ras/ERK through different mechanisms: these include dephosphorylation of RasGAP binding sites on specific receptors and adaptor proteins or dephosphorylation of the negative Ras/ERK regulator, Sprouty. Alternatively, regulation of Src kinase activity by Shp2 through either direct dephosphorylation of Src or through indirect regulation of the Src inhibitor Csk can enhance activation of the ERK pathway. Shp2 has also been described to regulate PI3K, Fak, and the Rho family GTPase, RhoA, as well as Ca^{2+} oscillations and Ca^{2+}/Calcineurin/NFAT signaling. In addition to RTK signaling, Shp2 has been implicated downstream of cytokine signaling in the regulation of Jak/Stat signaling pathways and in the activation of NF-κB (see references in the main text). Activation is indicated by arrows, inhibition by black circles, activating and inhibitory relationships by black squares, direct interactions by solid lines, and indirect interactions by dashed lines.

Stat proteins or by modulating the activity of the upstream Jak family kinases (Ali *et al.*, 2003; Chen *et al.*, 2003; Wu *et al.*, 2002). However, positive effects of Shp2 on Stat5 were also reported in mammary gland cells in culture and *in vivo* (Berchtold *et al.*, 1998; Chughtai *et al.*, 2002; Ke *et al.*, 2006). In addition, receptor- and/or cell context-dependent effects of Shp2 have been described on the activation of the phosphatidyl-inositol-3 kinase (PI3K)/Akt pathway (Ivins Zito *et al.*, 2004; Zhang *et al.*, 2002), of focal adhesion kinase (FAK) (Oh *et al.*, 1999; Yu *et al.*, 1998), and of Rho family GTPases (Kontaridis *et al.*, 2004; Schoenwaelder *et al.*, 2000). Shp2 has also been implicated in JNK (Shi *et al.*, 1998), NF-κB (You *et al.*, 2001), and NFAT (Uhlen *et al.*, 2006) activation in various settings (Fig. 2; reviewed in Neel *et al.*, 2003; Rosário and Birchmeier, 2003).

III. LESSONS FROM HUMAN AND MURINE GAIN-OF-FUNCTION MUTATIONS OF Shp2

A. Shp2 Gain-of-Function Mutations and Noonan Syndrome

The first human gain-of-function (GOF) mutation in Shp2 was discovered at the beginning of this decade in NS patients (Tartaglia *et al.*, 2001). NS is an autosomal dominant disorder affecting 1 in 1000–2500 live births (Opitz, 1985; Tartaglia *et al.*, 2001), although miscarriages do occur and the actual rate may be higher (reviewed in Tartaglia and Gelb, 2005). It is associated with craniofacial abnormalities, cardiac defects, short stature, and learning disabilities. There is, however, a large variability in the symptoms displayed in individuals with this syndrome. Craniofacial defects typically involve webbing of the neck, triangular faces due to an increased distance between eyes (hypertelorism) and oral malformations such as arched palate (Noonan, 1968; Noonan and Ehmke, 1963). Cardiac defects are a very important aspect of NS, and indeed NS is the most common cause of heart disease caused by mutations. Changes in heart morphology include most commonly pulmonic stenosis resulting from dysplastic valve leaflets, hypertrophic cardiomyopathy, and various septal defects (Marino *et al.*, 1999). Other symptoms variably present in individuals with NS include skeletal defects, in particular pectus excavatum, hearing loss, bleeding problems, and a general developmental delay (Tartaglia and Gelb, 2005).

Forty to 50% of NS patients carry mutations in the Shp2 gene, PTPN11 (Musante *et al.*, 2003; Tartaglia *et al.*, 2002; Yoshida *et al.*, 2004; Zenker *et al.*, 2004). Most mutations are missense mutations, with the N308D mutation in the PTP domain accounting for nearly 25% of the cases. Most

NS mutations affect amino acid residues that reside at or are close to the interface between the N-SH2 and PTP domains and therefore interfere with the autoinhibitory state of Shp2 (Fig. 1B). These mutant proteins promote the sustained activation of the ERK MAP kinases in transfected cells and in animal models (Araki et al., 2004; Chan et al., 2005; Fragale et al., 2004; Krenz et al., 2008; Mohi et al., 2005; Schubbert et al., 2005; Yu et al., 2006). The deregulated MAP kinase activity is most likely the underlying cause of the developmental dysfunctions seen in NS patients. Indeed, mutations in other components of the Ras/ERK pathway, such as KRAS, Raf, and SOS1, have been identified in NS patients lacking Shp2 mutations (Carta et al., 2006; Roberts et al., 2007; Schubbert et al., 2006; Tartaglia et al., 2007).

Mutations in Shp2 at homologous positions as those observed in NS patients were recently introduced into mice and flies. The phenotype in mice resembles to a large extent the one observed in humans (Table I; Fig. 3). Like the patients, Shp2 GOF (Shp2$^{D61G/+}$) mice show decreased body size, cardiac malformations, and reduced skull length, with the mice tending to display a triangular face, manifested by a wider and blunter snout (Araki et al., 2004). Hyperactivation of ERK1/2 signaling plays an important role in the development of craniofacial abnormalities as inhibition of ERK1/2 rescues skull defects of mice expressing the NS mutation Shp2^{Q79R} in the neural crest (Nakamura et al., 2007). An overview of mouse mutants generated in several laboratories is shown in Table I.

In the fly, ubiquitous heterozygous expression of the Shp2 homologue, Csw, bearing the N308D NS mutation has confirmed the GOF nature of this mutation and the importance of Shp2 for EGF signaling. N308D Csw transgenic flies show elevated ERK activation and a GOF EGF receptor-like phenotype: the induction of ectopic wing veins. As in the mouse (Zhang et al., 2009b), genetic crosses of N308D Csw expressing flies with mutant alleles for genes in the Jak/Stat pathway suggest that regulation of the Jak/Stat pathway by NS Csw also contributes to the phenotype (Oishi et al., 2006, 2009; Pagani et al., 2009).

B. Shp2 Gain-of-Function Mutations in Cancer

Somatic GOF mutations in the Shp2 gene, PTPN11, are found in about 35% of patients with sporadic JMML (Loh et al., 2004a; Tartaglia et al., 2005). JMML is a rare myeloproliferative and myelodysplastic disorder of young children that features massive expansion and tissue infiltration of myeloid cells and macrocytic anemia (Emanuel et al., 1996). As in NS, the mutations associated with JMML are located mainly in the N-SH2 domain of Shp2 and are GOF mutations that interfere with the autoinhibitory

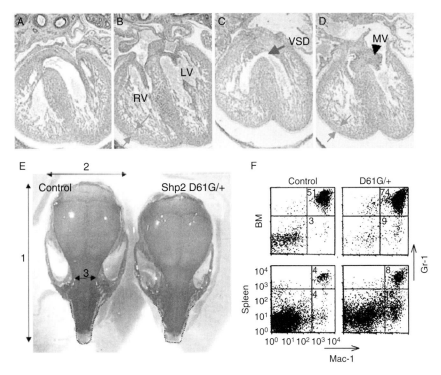

Fig. 3 The Noonan syndrome mouse, D61G. The D61G Noonan syndrome mouse develops heart and skull defects as well as myoproliferative disorder. Transverse sections through E13.5 hearts of control (A, B) and Shp2 D61G/+ mutant embryos (C, D) illustrate multiple heart defects in Shp2 D61G/+ mutants. These include ventricular septal defect (VSD, blue arrow) and hypertrophy of mitral valve (MV, black arrowhead) as well as thinner ventricular walls (red arrows). (E) Surviving Shp2 D61G/+ mutants exhibit facial dysmorphia similar to Noonan syndrome patients, which includes reduced skull length (1), normal skull widths (2), and reduced inner canthal distance (3). (F) Shp2 D61G/+ mice develop myeloproliferative syndrome that is confirmed by flow cytometric analysis showing myeloid expansion in bone marrow and spleen. From Araki et al. (2004).

interface of Shp2 (Fig. 1B). Indeed, based on structural information, Neel and collaborators had already postulated and shown that mutation of two N-SH2 domain residues, D61 and E76, that were subsequently shown to be mutated in JMML, results in increased phosphatase activity *in vitro* and *in vivo* in *Xenopus* ectodermal explants (O'Reilly et al., 2000). This and subsequent studies in animal cell lines, confirmed the GOF nature of these mutations and indicated that JMML mutations are more strongly activating than NS mutations (Keilhack et al., 2005; Niihori et al., 2005; Tartaglia et al., 2003). Indeed, although NS patients do have an increased incidence of JMML, JMML is not a hallmark of NS (Choong et al., 1999). Genetic abnormalities in three genes, PTPN11, Ras, and the Ras GAP, NF1, account for up to 85% of cases of JMML. This suggests that, as for NS, excessive Ras/ERK MAP kinase activation is important for the development of JMML (Loh *et al.*, 2004b; Tartaglia *et al.*, 2003).

Mice expressing a JMML-linked GOF mutation of Shp2 have been generated to address the role of Shp2 in tumorigenesis (Chan *et al.*, 2009). Mx1-Cre was used to induce expression of the Shp2^{D61Y} allele in 3-week-old mice. This resulted in lethality around 45 weeks after induction. Mutant mice developed myeloproliferative disorder characterized by myeloid expansion, elevated levels of myeloid precursors in the spleen, and tissue infiltration by granulocytes and macrophages. Nearly all animals developed anemia, a hallmark of JMML. Moreover, the hematopoietic Lin-Sca1+cKit+ stem cell compartment was expanded and showed increased stem cell factor-induced ERK and Akt activation and colony formation. Myeloid colony growth in the absence of exogenous cytokine stimulation and the selective hypersensitivity of bone marrow cells to GM-CSF are hallmarks of JMML in humans (Altman *et al.*, 1974; Emanuel *et al.*, 1991). Common myeloid progenitors and granulocyte–monocyte progenitors from Mx1-Cre Shp2$^{D61Y/+}$ mice, also produced cytokine-independent colonies in a cell-autonomous manner and displayed elevated ERK and Stat5 activation in response to GM-CSF. Thus, mice that carry a JMML-like Shp2 mutation develop myeloproliferative disorders that resemble those observed in patients.

Mice expressing a weaker NS and JMML-associated Shp2 mutation, D61G, also developed mild myeloproliferative disorder by 5 months of age (Araki *et al.*, 2004). Consistent with a cell-autonomous signaling defect in hematopoietic progenitors, bone marrow from Shp$^{D61G+/-}$ mice yielded factor-independent colonies, and hematopoietic progenitors from these mice showed increased sensitivity to IL-3 and GM-CSF (Araki *et al.*, 2004). Moreover, *in vitro* studies using expression of JMML-associated Shp2 mutations (D61Y and E76K) in mouse bone marrow-derived hematopoietic progenitors promoted both cell-cycle progression of these cells in response to low GM-CSF as well as their survival in minimal media conditions. As before, this was associated with increased ERK and Akt signaling, as well as changes in the expression of cell cycle and apoptosis regulators (Yang *et al.*, 2008).

PTPN11 mutations also occur at lower incidence in other myeloid neoplasms, such as acute myelogenous leukemia (AML), chronic myelomonocytic leukemia (CMML), and myelodysplastic syndrome (MDS) (Loh *et al.*, 2004a,b; Tartaglia *et al.*, 2003), as well as in B-acute lymphoblastic leukemia (B-ALL) (Tartaglia *et al.*, 2004; Yamamoto *et al.*, 2006). In addition, PTPN11 mutations have also been identified at low incidence in some solid tumors. These included neuroblastomas as well as lung carcinomas which are also associated with mutations in other Ras/ERK MAP kinase pathway components (Bentires-Alj *et al.*, 2004; Cotton and Williams, 1995; Ijiri *et al.*, 2000; Lopez-Miranda *et al.*, 1997; Martinelli *et al.*, 2006; Miyamoto *et al.*, 2008). Deregulated Shp2 activation may thus play a broad role in cancer pathogenesis.

IV. LESSONS FROM LOSS-OF-FUNCTION MUTATIONS OF Shp2

A. Shp2 LOF Mutations in Invertebrates and in *Xenopus*

The first evidence for a role of Shp2 in development came from LOF studies in invertebrates performed in the laboratories of Simon and of Perrimon (Freeman *et al.*, 1992; Herbst *et al.*, 1996; Perkins *et al.*, 1992, 1996). These studies were also pivotal in establishing Shp2 as an essential positive regulator of the Ras/MAP kinase pathway. In *Drosophila*, cell fate choices at the anterior and posterior embryonic termini require signaling by the PDGF receptor-like RTK, Torso. Corkscrew, the *Drosophila* Shp2 homologue, is required for the transduction of Torso signaling in this system (Perkins *et al.*, 1992, 1996). The effects of a Torso GOF mutation were suppressed by LOF mutations in Csw, and further epistasis experiments indicated that Csw worked downstream of Torso in concert with D-raf to control expression of terminal genes and the formation of terminal embryonic structures (Perkins *et al.*, 1992). Subsequently, Csw was found to be essential for signaling of another *Drosophila* tyrosine kinase receptor: Sevenless, the homologue of the vertebrate ros receptor. Sevenless signaling is required for the development of the R7 photoreceptor cell in the ommatidia of *Drosophila* eyes (Allard *et al.*, 1996; Herbst *et al.*, 1996). Ectopic expression of a membrane-targeted form of Csw was sufficient to rescue R7 photoreceptor development in the absence of a functional Sevenless. Likewise, dominant negative forms of Csw suppressed the function of activated Ras1 or Raf, suggesting that Csw acts either downstream of Ras1 or in parallel to Ras1/Raf during the determination of the R7 photoreceptor fate (Allard *et al.*, 1996; Herbst *et al.*, 1996).

Similarly, the function of *Caenorhabditis elegans* ptp-2 gene, a likely homologue of Shp2, is linked to the Ras/MAP kinase pathway. Ptp-2 LOF mutations that delete the PTP domain resulted in defects in oogenesis that could be rescued by GOF Ras (let-60), indicating that Ras acts downstream or in parallel to Ptp-2 (Gutch *et al.*, 1998). Shp2 is also required for MAP kinase activation during mesoderm induction in *Xenopus*. Microinjection of a catalytically inactive variant of Shp2 mRNA in *Xenopus* embryos caused severe posterior truncations, which were attributed to a deficit in the bFGF-stimulated activation of MAP kinase (Tang et al., 1995).

B. Null Mutations of Shp2 in Mice

The first targeted mutations of Shp2 in mice, $Shp2^{X3}$ and $Shp2^{X2}$, were reported in the mid-1990s and were apparently hypomorphic (Arrandale et al., 1996; Saxton et al., 1997). Splicing around exons 2 and 3 resulted in the expression of mutant Shp2 proteins that lacked most of the N-SH2 domain, but retained the remaining domains (Saxton et al., 1997; Yang et al., 2006). These mutations caused lethality at mid-gestation (Saxton et al., 1997). Neel and collaborators generated the first true Shp2 null allele by introducing, in addition to a deletion of exon 2, a splice acceptor site and sequences encoding β-galactosidase that "captured" initiated transcripts (Fig. 4, Table I; Yang et al., 2006). Embryos homozygous for this allele did not produce residual Shp2 protein and died at time of implantation, that is, much earlier than those carrying the hypomorphic alleles. Shp2 null blastocysts initially developed normally, but subsequently exhibited inner cell mass death, diminished numbers of trophoblast giant cells, and failure to yield trophoblast stem cell lines.

The first cell lineages established in the mammalian embryo are trophectoderm and inner cell mass (reviewed in Cross, 2000; Rossant and Cross, 2001). The inner cell mass generates the embryo proper, while the trophectoderm cells contribute to the extraembryonic tissues. The stem cell population of the trophectoderm lineage, trophoblast stem cells, can be isolated and maintained in the presence of FGF4 (Tanaka et al., 1998) and they can generate multiple trophoblast cell types in vitro and in vivo (Goldin and Papaioannou, 2003; Rossant, 2001; Tanaka et al., 1998). FGF4 signaling is essential for trophoblast development, as evidenced from the analyses of embryos that lack FGF4, FGFR2, FRS2a, and Erk2 (Arman et al., 1998; Feldman et al., 1995; Gotoh et al., 2005; Saba-El-Leil et al., 2003). In Shp2 LOF mutants, the trophoblast lineage was specified but failed to expand, and Shp2 mutant blastocysts did not yield trophoblast stem cell lines when cultured. Moreover, deletion of Shp2 in cultured trophoblast stem cells using cre-loxP technology ($Shp2^{flox/X11}$; Fig. 4) resulted in rapid apoptosis. This was attributed to a requirement of Shp2 for FGF4-evoked activation of the Ras/ERK pathway and thereby for destabilization of the proapoptotic protein, Bim, in these cells (Yang et al., 2006). Specifically, activation of FGF receptors by FGF4 binding results in the recruitment and phosphorylation of the adapter protein FRS2 that in turn recruits and activates Shp2. Shp2 then leads to the activation of src family kinases (SFK), possibly by regulating Csk recruitment, and hence activation of the ERK MAP kinase pathway. Finally,

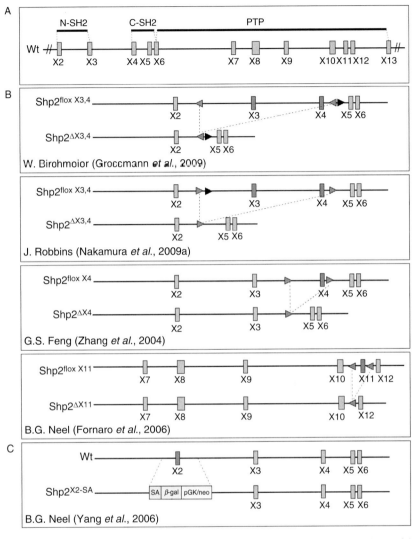

Fig. 4 Introduction of loss-of-function mutations into the Shp2 locus. The genomic locus of the Shp2 gene and its protein structure are shown in (A). Several strategies were used to generate Shp2 conditional loss-of-function alleles (B). In one strategy, loxP sites flank exons 3 and 4 to generate the Shp2flox X3, 4 allele. Exons 3 and 4 encode for part of the N-SH2 and part of the C-SH2 domain. In a second strategy, loxP sites flank exon 4 to generate the Shp2 flox X4 allele. In a third strategy, loxP sites flank exon 11 located in the PTP domain and generate the Shp2 flox X11 allele. After tissue-specific cre expression and excision of the floxed exons, all strategies lead to a frame shift, an early stop in protein translation and production of a null allele. For the generation of the Shp2 knockout allele, exon 2 encoding part of the N-SH2 domain was replaced by a cassette expressing β-gal/pGK-neo preceding a strong splice acceptor site that captures initiated transcript (C). The research laboratories that generated the specific mutated alleles and the first publications are indicated. LoxP and FRT sites are indicated by red and black triangles, respectively.

ERK phosphorylates and destabilizes Bim, promoting cell survival. The trophoblast stem cell is thus an example of transduction of an antiapoptotic growth factor signal through Shp2.

Shp2 also plays essential roles in embryonic stem cells. Feng and collaborators recently showed that homozygous deletion of Shp2 interferes with the differentiation of murine embryonic stem cells into cell types of all three germ layers. A similar block in differentiation of human embryonic stem cells was observed when Shp2 activity was reduced either through siRNA knockdown or using a small molecular weight inhibitor (Wu et al., 2009). Thus, Shp2 is crucial for the maintenance of pluripotency in cultured embryonic stem cells. Interestingly, the effects observed in human and murine embryonic stem cells were somewhat different. Whereas in both cell types ERK activity was reduced upon downregulation of Shp2, the effects on Stat3 and Smad 1/5/8 diverged. Thus, common (ERK) and distinct pathways (Stat, BMP) appear to be controlled by Shp2 in the human and murine embryonic stem cells.

C. LOF Mutations of Shp2 in Humans: LEOPARD Syndrome

Heterozygous germline mutations in PTPN11 have been found in more than 80% of LEOPARD syndrome cases (Digilio et al., 2004; Keren et al., 2004). LEOPARD syndrome is a rare autosomal dominant disorder characterized by *L*entigenes (scattered lentil-shaped pigmentation, but also café au lait spots), *E*CG abnormalities, *O*cular hypertelorism, *P*ulmonic valvular stenosis, *A*bnormalities of genitalia, *R*etardation of growth, and *D*eafness. The syndrome thus shares many similar symptoms to NS. Lentigenes are however rare in NS and webbed neck, skeletal defects, and bleeding disorders are absent in LEOPARD syndrome (Noonan, 2006; Ogata and Yoshida, 2005; Tartaglia and Gelb, 2005).

Germline PTPN11 mutations in LEOPARD syndrome all reside in the PTP domain. Unlike NS mutations though, LEOPARD syndrome mutations affect residues of the PTP domain involved in catalysis and are predicted to be LOF mutations. Indeed, biochemical evidence suggests that LEOPARD syndrome alleles inhibit growth factor-stimulated ERK activation in transient and stable transfection assays (Hanna et al., 2006; Kontaridis et al., 2006; Tartaglia et al., 2006). LEOPARD syndrome mutations in Shp2 may thus result in an open conformation, inactive Shp2 molecule that may in addition act as a dominant negative and interfere with the function of wild-type Shp2 (Kontaridis et al., 2006). Despite the opposing effects of Noonan and LEOPARD syndrome mutations on Shp2 activity, Noonan and LEOPARD syndrome patients exhibit similar heart defects, facial abnormalities, short stature, and increased incidence of tumors, although the type of tumors differ (Noonan, 2006; Ogata and Yoshida, 2005; Tartaglia and Gelb, 2005).

The similarities between Noonan and LEOPARD syndromes may be initially surprising but several possible explanations exist. Loss of one PTPN11 allele in the mouse has no phenotype, suggesting that the effect of the LEOPARD mutation on the *in vivo* signaling functions of Shp2 may be more complex than a simple LOF. Recent expression of common LEOPARD syndrome mutations (Y279C and T468M) in *Drosophila* has suggested that these mutants may also possess GOF activity during *Drosophila* development (Oishi *et al.*, 2009). *In vitro* assays demonstrated that introduction of these mutations into Csw significantly reduced phosphatase activity, as had been observed for mammalian Shp2. Nevertheless, ubiquitous expression of these LEOPARD-Csw mutants in *Drosophila* induced ectopic wing veins and a R7-receptor associated rough eye phenotype in a phosphatase-domain-dependent fashion. Both of these phenotypes are associated with activation of the Ras/MAP kinase pathway and are induced by expression of NS mutations in Csw. A caveat with these experiments, however, is that transgenic Csw expression in these flies is significantly higher than that of endogenous Csw.

An alternative explanation is that roles of Shp2 at different developmental stages, upstream of different effectors or downstream of different receptor systems may result in similar effects when aberrantly hyper- or hypoactivated. This is particularly true in the formation of a complex organ such as the heart, where tightly controlled sequential stages of proliferation, migration, cell-cycle arrest, and morphogenesis are required for successful organogenesis. In support of this theory, both GOF and LOF mutations in the MAP kinase activator, Raf1, have been identified in both Noonan and LEOPARD syndrome patients (Pandit *et al.*, 2007). The authors reported, however, a positive correlation between Raf1 kinase activity and the occurrence of hypertrophic cardiomyopathy in NS patients, suggesting tissue or cell type-specific effects of these mutations and emphasizing the complexity of these syndromes.

V. ORGAN SPECIFIC ROLES OF Shp2: LESSONS FROM MUTATIONS IN MICE

A. Shp2 and Heart Development

NS patients and corresponding mouse models, for example $Shp2^{D61G}$ or $Shp2^{N308D}$ mice, show cardiac malformations that include valvular stenosis, atrial and ventricular or atrioventricular septal defects, and double-outlet right ventricles (Table I, Fig. 3). In mice, the severity of heart phenotypes depends on the particular allele of Shp2, the gene dosage (heterozygous vs. homozygous mutations), and also on the genetic background. In severe cases, the malformations lead to embryonic lethality (Araki et al., 2009). It is interesting to note that genetic modifiers affect the severity of the heart malformations in mice; such modifiers might also explain the variability of the phenotypes observed in patients.

The vertebrate heart is generated from three different progenitor cell lineages, the mesoderm-derived first and second heart field cells and the cardiac neural crest cells (reviewed in Buckingham et al., 2005; Srivastava, 2006). First heart field cells form the left ventricle, while second heart field cells build the future right ventricle as well as inflow and outflow tracts. Cardiac neural crest cells contribute to the remodeling of arch arteries, septation of the outflow tract, closure of the ventricular septum, and innervation of cardiac ganglia (Jiang et al., 2000; Kirby and Waldo, 1990; Kirby et al., 1983). In order to understand the origin of the cardiac defects in Noonan mutants, conditional GOF Shp2 mutations, $Shp2^{D61Y}$, were produced in mice by Neel and collaborators, using αMHC-, Tie2-, or Wnt1-cre recombinase lines, which induced the mutations in cardiomyocytes, endothelial, or neural crest-derived cells, respectively (Araki et al., 2009). Initially, it was believed that mutation of Shp2 only in endocardial cells reproduced the heart phenotype of NS (Araki et al., 2004, 2009). However, a recent study by Robbins and collaborators showed that transgenic mice expressing the activating mutation, $Shp2^{Q79R}$, under the cardiomyocyte-specific βMHC promoter, displayed cardiac defects such as ventricular noncompaction and septal defects (Nakamura et al., 2007). In addition, overexpression of $Shp2^{Q79R}$ in the endothelial-derived cell lineage using Tie2-cre was shown to lead to enlarged endocardial cushions caused by increased cell proliferation and reduced apoptosis within the cushion mesenchyme and endothelium (Krenz et al., 2008). It is therefore likely that deregulation of Shp2 in both cardiomyocyte and endothelial cells contribute to the cardiac defects seen in NS patients. These NS mutations induced by MHC-cre (for cardiomyocytes) and Tie2-cre (for endothelial cells) both resulted in increased activation of ERK1/2 (Araki et al., 2009; Krenz et al., 2008; Nakamura et al., 2007). Furthermore, genetic modulation of ERK 1/2 expression rescued both endocardial cushion enlargement induced by Tie2-cre $Shp2^{Q79R}$ expression and ventricle noncompaction as well as heart function in β-MHC-cre $Shp2^{Q79R}$ mice (Krenz et al., 2008; Nakamura et al., 2007). Transgenic overexpression of ERK1/2 also phenocopied the NS phenotype in the heart. These studies have stressed the importance of the Ras/ERK MAP kinase signaling downstream of Shp2 in myocardial and endocardial tissue for normal heart development (Krenz et al., 2008).

LOF mutations of Shp2 have also been generated in the mouse heart. Loss of Shp2 expression in the heart muscle using either αMHC-cre or Mck-cre lines leads to dilated cardiomyopathy associated with defective Ras/ERK activation (Fig. 5; Kontaridis *et al.*, 2008; Princen *et al.*, 2009). In addition, Shp2-defective cardiomyocytes also show hyperactivation of the small GTPase RhoA which may contribute to abnormal cardiomyocyte morphology (Kontaridis *et al.*, 2008). The Mck-cre is expressed both in cardiomyocytes and skeletal muscle. In this mouse model, dilated cardiomyopathy is coupled with insulin resistance, glucose intolerance, and impaired glucose uptake in cardiomyocytes and striated muscle cells (Princen *et al.*, 2009). The authors report a variety of alterations in downstream signaling in these

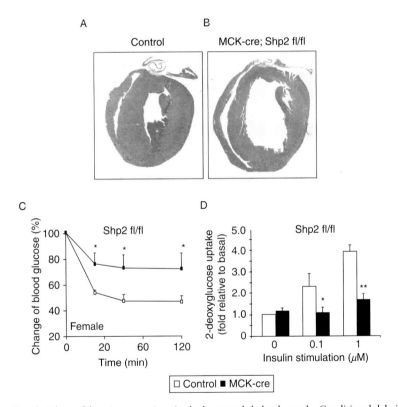

Fig. 5 Shp2 loss-of-function mutations in the heart and skeletal muscle. Conditional deletion of Shp2 using a myosin-specific cre results in heart and muscle defects as well as metabolic changes. Deletion of Shp2 in cardiomyocytes using the MCK-cre line causes dilated cardiomyopathy, here shown in hearts from 8-week-old control (A) and mutant (B) mice. MCK-cre Shp2 fl/fl mutant mice also exhibit impaired insulin sensitivity and glucose intolerance (C, D). From Kontaridis *et al.* (2008) and Princen *et al.* (2009).

two cell types including changes in LIF-induced activation of PI3K, ERK5, and Stat3 pathways in cardiomyocytes as well as impaired protein kinase C-ζ/λ and AMP-kinase activities in striated muscle (Princen et al., 2009). A further study used the same Mck-cre, as well as in vitro transfection of Cre into cultured floxed Shp2 myoblasts, to study the role of Shp2 during skeletal muscle development (Fornaro et al., 2006). Here, defects in myotube formation and the associated reduction in myofiber size and type I slow myofiber numbers upon loss of Shp2 expression were attributed to defective NFAT signaling. Both NFAT4 and NFAT1 have been shown to influence muscle mass by controlling myofiber number (Kegley et al., 2001) and myofibers size (Horsley et al., 2001), respectively. Thus, Shp2 serves a critical role in cardioprotection and adult cardiac and skeletal muscle function. Conditional LOF mutations of Shp2 in the neural crest that affect the heart are described in Section D below.

B. Shp2 in T-Cell Development

The early studies using the hypomorphic Shp2^{X3} allele in mice had suggested a function of Shp2 in erythroid and myeloid cell lineage commitment as well as in T and B cell development (Qu et al., 1997, 1998, 2001). Intrathymic T-cell-lymphocyte differentiation comprises a strictly regulated developmental program, which is initiated after fetal liver or bone marrow-derived precursor cells enter the thymus. Thymocyte maturation involves a series of stages in which signals provided by growth factors, adhesion molecules, and T-cell antigen receptors (TCR) dictate, which cells will exit the thymus as mature T lymphocytes. These stages can be readily followed by the ordered expression of a number of cell-surface markers. In order to elucidate the function of Shp2 during T-cell development, Feng and collaborators made use of the Lck-cre, which is expressed in T-cell during development (Nguyen et al., 2006); Lck is a cytoplasmic tyrosine kinase involved in thymocyte development (Molina et al., 1992). By conditional deletion of Shp2 in the thymus, it was shown that Shp2 controls T-cell differentiation/proliferation instructed by the pre-TCR. The Shp2 deletion significantly suppressed expansion of CD4$^+$ T cells. Furthermore, Shp2 also acts as a positive regulator of TCR signaling in mature T-lymphocytes. The reported phenotype changes were attributed to reduced phosphorylation of ERK1/2 in Shp2 mutant thymocytes and are consistent with an essential function of Ras/ERK signaling during thymocyte development (Alberola-Ila et al., 1995; Crompton et al., 1996; Swat et al., 1996).

C. Shp2 in the CNS

NS is associated with mild mental retardation and learning disabilities such as hyperactivity disorder (Noonan and Ehmke, 1963; Opitz, 1985; Tartaglia and Gelb, 2005; Tartaglia et al., 2001). The cerebral cortex plays a key role in memory, perceptual attention, language, and consciousness. During the development of the cortex, tightly regulated asymmetric divisions of precursor cells lying in the ventricular zone give rise firstly to the different neuronal cell types and layers of the cortex, and subsequently to glial cells. The timing of the generation of these different cell types is tightly controlled and is surprisingly conserved in neuronal precursor cell cultures (reviewed in Miller and Gauthier, 2007). Thus, cortical neurogenesis occurs during embryogenesis, while gliogenesis occurs largely postnatally (Lillien, 1998; Shen et al., 1998). Although much is still unknown as to how this "neurogenic-to-gliogenic switch" is controlled, RTK signaling pathways and activated ERK-C/EBP pathway have been shown to be required for neurogenesis (Menard et al., 2002; Park et al., 1999; Raballo et al., 2000; Williams et al., 1997) while other pathways such as the gp130/Jak/Stat pathway are required for the later onset of astrogenesis (Bonni et al., 1997; Johe et al., 1996; Nakashima et al., 1999; Sun et al., 2001). Expression of GOF Shp2 mutants in the cortex, through the generation of Shp2$^{D61G/+}$ mice or through *in utero* electroporation of Shp2^{D61G}, perturbs the development of this organ (Gauthier et al., 2007). This has been attributed to an alteration in the balance between neurogenesis and astrogenesis during development of this organ, with increased number of neurons and decreased number of glial cells being formed. Indeed, the opposite result was observed upon interference with Shp2 expression by either conditional deletion of Shp2 in neural progenitor cells using Nestin-Cre (Ke et al., 2007) or using an shRNA-mediated knockdown of Shp2 in cortical precursors in culture or *in utero* (Gauthier et al., 2007). Interference with Shp2 expression in neural progenitor cells resulted in impaired corticogenesis associated with reduced proliferation of progenitor cells, decreased neurogenesis, and increased gliogenesis (Gauthier et al., 2007; Ke et al., 2007). This differential regulation of neurogenesis and gliogenesis by Shp2 was proposed to occur through the Shp2-mediated promotion of the neurogenic ERK MAP kinase pathway and inhibition of the gliogenic Stat3 pathway.

Activation of the ERK MAP kinase pathway downstream of bFGF is also required for the self-renewing proliferation of neural progenitor cells. bFGF activates the ERK pathway by recruiting Shp2 through association with the adaptor protein FRS2α (Hadari et al., 1998). Indeed, FRS2α knockin mice (FRS2$^{\Delta Shp2/\Delta Shp2}$) that lack the two Shp2 binding sites display severely impaired cortical development and neurogenesis, in part due to defects in intermediate progenitor cells (Yamamoto et al., 2005). Shp2-deficient neural progenitor cells also show defective proliferation and self-renewal (Ke *et al.*, 2007). Defective activation of the ERK pathway downstream of bFGF in these cells, and the subsequent decreased phosphorylation and activation of transcription factors such as c-Myc, probably accounts for the impaired self-renewal proliferation of Shp2-deficient neural progenitor cells. Indeed, transfection of the c-Myc target Bmi-1, a polycomb family transcriptional repressor, increased proliferation and second neurosphere formation in Shp2-deficient neural progenitor cells (Ke *et al.*, 2007).

Conditional Shp2 null mutations in the mouse have also demonstrated roles for Shp2 during cerebellar development. Nestin-cre Shp2 mutant mice displayed reduced and less foliated cerebellum, mispositioned Purkinje cells, and ectopic presence of external granule cells. In the latter case, the phenotype was attributed to defects in SDF-1 or CXCR4 induced granule cell motility (Hagihara *et al.*, 2009).

Shp2 may also play important roles in the adult brain. Signaling by the leptin receptor, a cytokine receptor of the gp130 family, is involved in regulating energy balance and metabolism in the adult (Zhang *et al.*, 2004). Leptin is secreted by adipocytes and regulates the size of adipose tissue mass through effects on satiety and energy metabolism by binding to and activating the leptin receptor in the hypothalamus. Obesity is a strong predisposing factor for type 2 diabetes and cardiovascular disease. Using the CaMKII-Cre, Shp2 was deleted postnatally in neuronal cells of the cerebral cortex and hypothalamus (Table I). Mutant mice developed early-onset obesity and diabetes that was accompanied by increased levels of leptin, insulin, glucose, and triglycerides in their blood. Mutant mice showed altered Leptin signaling in the hypothalamus which included decreased activation of ERK2 and mild activation of Stat3 (Zhang *et al.*, 2004). Leptin signaling has been shown to activate the Jak/Stat pathway, and disruption of Leptin receptor or of Stat signaling in the mouse also leads to obesity (Bates *et al.*, 2003; Gao *et al.*, 2004). This suggests that other positive effects of Shp2, perhaps through ERK2, on Leptin signaling outweigh its negative effect on Stat signaling in the development of the obesity phenotype. Deletion of Shp2 in the mouse using the pan-neuronal cre line Cre3 also led to early development of obesity and diabetes as well as a number of complications commonly associated with these conditions in humans (Krajewska *et al.*, 2008).

Specific genetic ablation or activation of Shp2 in the mouse brain has thus demonstrated roles for Shp2 during both early lineage decisions and cell migrations during development as well as in the regulation of energy metabolism in the adult. Although in each case, Shp2 appears to function downstream of distinct receptor systems, activation of the ERK MAP kinase pathway downstream of Shp2 is a common and necessary event for correct organ formation and function.

D. Shp2 and Neural Crest-Derived Tissues

Neural crest cells constitute a transient population of multipotent stem cells that emerges from the neural tube and generates many different cell types (Christiansen *et al.*, 2000; Crane and Trainor, 2006; Le Douarin *et al.*, 2004). The cell type formed depends on the rostrocaudal origin of the neural crest cells, and on the target site at which they settle. Thus, cranial neural crest cells contribute to the craniofacial skeleton and cranial sensory ganglia, vagal, and sacral neural crest cells contribute to the heart and give rise to the parasympathetic and enteric nervous system, whereas trunk neural crest cells generate melanocytes, chromaffin cells of the adrenal gland, sensory and sympathetic neurons, and the glial cells associated with such neurons.

Shp2 is essential for the development of several neural crest cell derivatives, for instance for the generation and differentiation of Schwann cells (Grossmann *et al.*, 2009; Nakamura *et al.*, 2009a,b). Schwann cells are neural crest-derived glial cells, whose precursors accompany developing peripheral nerves. In the perinatal period, Schwann cells undergo terminal differentiation, that is, they ensheath nerves and begin to elaborate the complex myelin sheath that allows the fast conduction of nerve impulses (saltatory conduction) (Birchmeier, 2009; Jessen and Mirsky, 2005; Lemke, 2006). Schwann cell precursors that associate with peripheral axons express the tyrosine kinase receptors ErbB2/3, and depend on these receptors and on Neuregulin-1 (Nrg1), the neuronally produced ligand, for migration, proliferation, and differentiation (Crane and Trainor, 2006; Dong *et al.*, 1995; Lyons *et al.*, 2005; Meyer and Birchmeier, 1995; Michailov *et al.*, 2004; Morris *et al.*, 1999; Riethmacher *et al.*, 1997; Taveggia *et al.*, 2005; Woldeyesus *et al.*, 1999). Conditional mutation of Shp2 generated in our laboratory using Wnt1-cre results in a pronounced reduction in the numbers of Schwann cell precursors, similar to what was observed in Nrg1, ErbB2, or ErbB3 mutant mice (Fig. 6). Our biochemical analysis revealed a cell-autonomous function of Shp2 in Schwann cells. Cells mutant for Shp2 were unable to migrate or proliferate in response to Nrg1 in culture. Several aspects of the Nrg1-dependent signaling were affected in mutant Schwann cells, namely sustained activation of ERK1/2 and activation of Src and Fak. Pharmacological inhibition of Src family kinases mimicked all cellular and biochemical effects of the Shp2 mutation, implicating Src as a primary Shp2 target during Nrg1 signaling. In addition, Shp2 is essential for Schwann cell myelination, as revealed by the conditional mutation of Shp2 late in Schwann cell development using Krox-20-cre (Grossmann *et al.*, 2009). Mutant mice displayed strong hypomyelination similar to what was observed in Krox20-cre ErbB2 mutant mice (Garratt *et al.*, 2000). Moreover, a large overlap in the changes in gene expression between the Shp2 and ErbB2 mutant peripheral nerves was shown by the use of Affymetrix

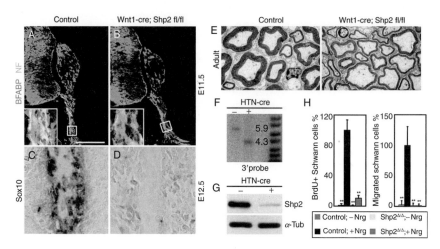

Fig. 6 Shp2 is essential for Schwann cell development and myelination. Shp2 is required for several facets of Schwann cell development and differentiation. Conditional deletion of Shp2 in Schwann cell precursors early during development, using the Wnt1-cre, affects Schwann cell migration with the projecting nerves. This is illustrated here at E11.5 by loss of BFABP positive Schwann cell precursors associated with peripheral nerves (stained for NF) projecting from the spinal cord and dorsal root ganglia (A, B) and, at E12.5, by loss of Sox10 positive glial cells associated with these axon tracts (C, D). Loss of Shp2 expression later in development through Krox20-cre-mediated recombination, on the other hand, interferes with myelination so that Shp2 mutant mice have only thin myelin sheaths (E). Southern blot and Western blot analyses of Shp2 after HTN-cre induced recombination in cultured Shp2fl/fl Schwann cells (F, G). At a cellular level, we could show that Shp2 is required for the proliferation and migration of Schwann cells in response to neuregulin. This is illustrated here in cultured floxed Shp2 Schwann cells treated with membrane-penetrable Tat-cre (HNT-cre) (H). From Grossmann et al. (2009).

microarray analyses. Together, these genetic and biochemical analyses have demonstrated that Schwann cells depend on Shp2 for the transduction of Nrg1/ErbB signals (Grossmann et al., 2009). Additional developmental alterations are apparent in these Wnt1-cre mutant mice, for instance in the enteric nervous system, melanocytes, or in craniofacial structures (see also below).

Robbins and collaborators recently analyzed in detail the role of Shp2 in the development of the heart outflow tract and craniofacial structures, two tissues that are also generated by the neural crest (Nakamura et al., 2009a). In this report, LOF mutations in Shp2 were introduced by conditional mutation using Wnt1-cre (Fig. 4, Table I). Ablation of Shp2 led to heart defects such as failure of outflow tract septation (division into the pulmonary artery and aorta) and abnormalities of the great arteries, apparently caused by defective migration of cardiac neural crest cells into the proximal outflow tract of the developing heart by E12.5. Biochemical analyses revealed reduced phosphorylation of ERK1/2 in Shp2-deficient neural crest cells. Conditional inactivation of ERK1/2 in the developing neural crest led to similar changes, providing further genetic support for the role of Shp2 in the control of ERK1/2 activation (Newbern et al., 2008).

Cranial neural crest cells have the potential to differentiate into cartilage, bone, and connective tissue. Cranial paraxial mesoderm contributes to the skeletal muscle tissue of the face, and communication between cranial neural crest and cranial mesoderm provide a crucial driving force for craniofacial development. Wnt1cre-Shp2 mutant embryos have a smaller head, nose, jaw, and ear, as well as aberrant eye placement. At E17.5, large proportions of the face and cranium were affected in Wnt1cre-Shp2 mutants, including dramatic ablation of mandible and frontal and nasal bones, concomitant with loss of the osteoblast differentiation marker osteopontin (Nakamura et al., 2009a). Interestingly, mutation of ERK1/2 in neural crest cells leads to similar developmental defects (Newbern et al., 2008). Thus, many if not all types of neural crest cells are dependent on Shp2 for their development. It is interesting to note that most of these neural crest cell types depend on tyrosine kinase receptor for migration, growth, and differentiation, and that different neural crest cell population use distinct receptors such as c-kit, c-ret, erbB2/3, FGFRs for these functions. The broad role of Shp2 in neural crest cells probably thus reflects a requirement of Shp2 in signaling by different tyrosine kinase receptors.

E. Shp2 in Pancreas, the Liver, and Mammary Gland

Conditional mutations of Shp2 have also been produced in organs that contain epithelia, that is, the pancreas, liver, and mammary gland. The pancreas contains endocrine cells that produce hormones like insulin and glucagon, as well as exocrine cells that secrete digestive enzymes. Pancreatic differentiation begins with the endodermal expression of the homeodomain transcription factor Pdx1 that controls balanced differentiation of endocrine and exocrine progenitors. Moreover, Pdx1 plays a crucial role in the adult pancreas, where it participates in controlling the gene expression program of differentiated β-cells, for instance insulin expression. Feng and collaborators have produced mice with conditional ablation of Shp2 in the pancreas using Pdx1-cre (Fig. 4, Table I; Zhang et al., 2009a). These mice developed normally to adulthood, but displayed decreased insulin production and progressively impaired glucose tolerance. The decreased insulin production in Shp2-deficient β-cells was caused by impaired expression of the genes encoding insulin, Ins1 and Ins2. This was accompanied by a reduction in Pdx1 expression and activity, which together resulted in a dramatic decrease of Pdx1 binding to the Ins1 and Ins2 promoters. siRNA-mediated knockdown of Shp2 in the glucose-responsive insulinoma cell line INS-1 832/13 had revealed a role of Shp2 in activating the Akt/FoxO1 and ERK signaling cascades. Pdx1 is repressed by FoxO1, whose activity in turn is controlled by Akt through the Akt phosphorylation-dependent nuclear exclusion of FoxO1 (Kitamura *et al.*, 2002; Nakae *et al.*, 2002). ERK1/2, on the other hand, was shown to phosphorylate and activate Pdx1 and other transcription factors involved in the control of insulin promoters (Lawrence *et al.*, 2008; Ueki *et al.*, 2006). Mechanistically, Shp2 may control phosphoinositide-3 kinase (PI3K) and ERK signaling in β-cells by direct interaction with the insulin receptor substrate (IRS1 and family members), PI3K and with the negative regulator of the Ras/MAP kinase pathway, Sprouty1. Thus, in pancreatic β-cells Shp2 appears to act as a coordinator of several signaling pathways that control insulin synthesis and glucose homeostasis.

Hepatocytes are mostly quiescent, highly differentiated cells. Therefore, it is perhaps surprising that liver regeneration after most forms of injury does not rely on stem or progenitor cells, but instead involves proliferation of the differentiated hepatocytes (Fausto and Campbell, 2003). Cell-cycle entry of hepatocytes during regeneration involves upregulation of a number of cytokines, growth factors, and hormones as well as immediate early genes and cell-cycle regulators (Fausto *et al.*, 1995). In particular, hepatocyte growth factor/scatter factor (HGF/SF) and its receptor Met as well as interleukin-6 (IL-6) and gp130 play pivotal roles during the regeneration process (Borowiak *et al.*, 2004; Cressman *et al.*, 1996; Huh *et al.*, 2004; Wuestefeld *et al.*, 2003). Using the liver-specific Alb-cre, Feng and collaborators showed that both Shp2 and the adaptor protein Gab1 (Grb2-associated binder 1), serve crucial roles during liver regeneration (Bard-Chapeau *et al.*, 2005, 2006). Gab1 becomes tyrosine phosphorylated in response to stimulation by multiple growth factors, and thereby recruits and activates a number of signaling effectors, including Shp2 (Holgado-Madruga *et al.*, 1996; Maroun *et al.*, 2000; Montagner *et al.*, 2005; reviewed in Rosário and Birchmeier, 2003). The interaction between Gab1 and Shp2 is stimulated during liver regeneration (Bard-Chapeau *et al.*, 2006). Ablation of either Shp2 or Gab1 in the liver diminishes the regeneration process and is associated with reduced ERK activation and reduced expression of immediate early genes like c-fos and c-jun, or of cell-cycle regulators like cyclin A, E, and B1 (Bard-Chapeau *et al.*, 2005, 2006). In contrast, tyrosine phosphorylation of Stat3 and IL6 levels were increased in Gab1 or Shp2 mutant mice, possibly due to increased hepatic inflammation and necrosis. Taken together, these data demonstrate that both Gab1 and Shp2 play an essential role in mitogenic signaling during liver regeneration. Gab1 and Shp2 thus provide a link between receptors (met, gp130) and cell-cycle regulators during liver regeneration.

The mammary gland is built by two cell compartments: the epithelium, which consists of ducts and milk-producing alveolar cells, and the stroma or connective tissue, which is also called the mammary fat pad. During embryonic development, sequential and reciprocal signals between the epithelium and the stroma direct the outgrowth of a small duct into deeper layers of the dermal mesenchyme. During puberty, further elongation and bifurcation of the duct occurs so that in the mature animal, the entire fat pad is filled with a regularly spaced system of primary and secondary ducts that are decorated with side branches that form and disappear during each oestrous cycle. During pregnancy, alveolar proliferation and maturation occur, after which the gland reaches maturity and lactation is initiated (reviewed in Hennighausen and Robinson, 2005; Watson and Khaled, 2008). Mice deficient in Shp2 in the mammary gland were generated by conditional mutagenesis using MMTV-Cre transgenic mice (Ke *et al.*, 2006). These mice exhibited delayed and impaired morphogenesis of alveolar structures accompanied by reduced milk secretion. Deletion of Shp2 resulted in a reduction of Stat5 and ERK phosphorylation and activity, whereas Stat3 activity was somewhat upregulated. It was suggested that Shp2 is important for prolactin-induced formation of a trimeric complex that consists of the prolactin receptor (PrlR), Jak2, and Stat5 (Ali *et al.*, 1996). Genetic analysis demonstrated that in the absence of Stat5, the mammary epithelium develops ducts but fails to form alveoli and to produce milk protein (Chapman *et al.*, 1999; Liu *et al.*, 1997). Thus, while in other tissues a negative role of Shp2 in Stat3 signaling was observed, in the mammary gland Shp2 positively regulates Stat5 activation (Gauthier *et al.*, 2007; Ke *et al.*, 2007; Princen *et al.*, 2009).

Shp2 may exert these positive effects on JAK/Stat5 signaling by directly dephosphorylating JAK2 on a tyrosine residue required for interaction with SOCS-1, an ubiquitin ligase-associated adaptor protein, and thus promoting the stability of JAK2 (Ali *et al.*, 2003). Shp2 may also affect the stability of JAK2-association with PrlR through dephosphorylation of the SOCS docking protein.

VI. CONCLUSIONS AND PERSPECTIVES

The last two decades have seen an enormous expansion in our knowledge of function of the tyrosine phosphatase Shp2. Most importantly, GOF and LOF mutations of Shp2 were discovered in human patients, which lead to various developmental disturbances and cancer. Shp2 mutations with similar phenotypes were generated in mice and other vertebrates, but also in invertebrates like *Drosophila* and *C. elegans*. Overall, the results demonstrate that Shp2 acts in evolutionary conserved pathways, mostly downstream of RTKs and upstream of the Ras/ERK pathway. In these pathways, Shp2 has emerged as a positive, apparently nonredundant regulator of signaling.

Experimental mutations in mice have revealed essential Shp2 functions for ES cell differentiation and for the development of all three embryonic germ layers as well as for the subsequent development and function of specific cell lineages in the CNS, PNS, heart, in epithelial organs, and during T-cell development. In these tissues, Shp2 appears to regulate a diverse range of cellular functions including proliferation, differentiation, and migration, in a context-dependent fashion. The receptor system requiring Shp2 function also appear to be cell- and context-dependent. Shp2, however, does not seem to act in other signaling systems, where tyrosine phosphorylation and dephosphorylation play a role, suggesting that Shp2 activators are important in regulating specific activation of Shp2.

The wide variety of cellular and developmental roles that have been uncovered for Shp2 suggest regulation of a variety of different signaling pathways by this phophatase. Perhaps surprisingly then, regulation of the ERK MAP kinase pathway is a common theme in all of these tissues. Only a few pathways, such as the Stat and PI3K pathways, have so far been reported to be regulated by Shp2 in a cell-specific fashion. We also still lack much knowledge concerning the direct targets of Shp2 and thus, how particular downstream pathways are regulated. Further tissue- or event-specific signaling functions of Shp2 are sure to still be uncovered, and much still remains to be studied concerning the molecular and cellular function of this versatile protein.

REFERENCES

Adachi, M., Sekiya, M., Miyachi, T., Matsuno, K., Hinoda, Y., Imai, K., and Yachi, A. (1992). Molecular cloning of a novel protein-tyrosine phosphatase SH-PTP3 with sequence similarity to the src-homology region 2. *FEBS Lett.* **314**, 335–339.

Ahmad, S., Banville, D., Zhao, Z., Fischer, E. H., and Shen, S. H. (1993). A widely expressed human protein-tyrosine phosphatase containing src homology 2 domains. *Proc. Natl. Acad. Sci. USA* **90**, 2197–2201.

Alberola-Ila, J., Forbush, K. A., Seger, R., Krebs, E. G., and Perlmutter, R. M. (1995). Selective requirement for MAP kinase activation in thymocyte differentiation. *Nature* **373**, 620–623.

Ali, S., Chen, Z., Lebrun, J. J., Vogel, W., Kharitonenkov, A., Kelly, P. A., and Ullrich, A. (1996). PTP1D is a positive regulator of the prolactin signal leading to beta-casein promoter activation. *EMBO J.* **15**, 135–142.

Ali, S., Nouhi, Z., Chughtai, N., and Ali, S. (2003). SHP-2 regulates SOCS-1-mediated Janus kinase-2 ubiquitination/degradation downstream of the prolactin receptor. *J. Biol. Chem.* **278,** 52021–52031.

Allard, J. D., Chang, H. C., Herbst, R., McNeill, H., and Simon, M. A. (1996). The SH2-containing tyrosine phosphatase corkscrew is required during signaling by sevenless, Ras1 and Raf. *Development* **122,** 1137–1146.

Altman, A. J., Palmer, C. G., and Baehner, R. L. (1974). Juvenile "chronic granulocytic" leukemia: a panmyelopathy with prominent monocytic involvement and circulating monocyte colony-forming cells. *Blood* **43,** 341–350.

Araki, T., Nawa, H., and Neel, B. G. (2003). Tyrosyl phosphorylation of Shp2 is required for normal ERK activation in response to some, but not all, growth factors. *J. Biol. Chem.* **278,** 41677–41684.

Araki, T., Mohi, M. G., Ismat, F. A., Bronson, R. T., Williams, I. R., Kutok, J. L., Yang, W., Pao, L. I., Gilliland, D. G., Epstein, J. A., and Neel, B. G. (2004). Mouse model of Noonan syndrome reveals cell type- and gene dosage-dependent effects of Ptpn11 mutation. *Nat. Med.* **10,** 849–857.

Araki, T., Chan, G., Newbigging, S., Morikawa, L., Bronson, R. T., and Neel, B. G. (2009). Noonan syndrome cardiac defects are caused by PTPN11 acting in endocardium to enhance endocardial-mesenchymal transformation. *Proc. Natl. Acad. Sci. USA* **106,** 4736–4741.

Arman, E., Haffner-Krausz, R., Chen, Y., Heath, J. K., and Lonai, P. (1998). Targeted disruption of fibroblast growth factor (FGF) receptor 2 suggests a role for FGF signaling in pregastrulation mammalian development. *Proc. Natl. Acad. Sci. USA* **95,** 5082–5087.

Arrandale, J. M., Gore-Willse, A., Rocks, S., Ren, J. M., Zhu, J., Davis, A., Livingston, J. N., and Rabin, D. U. (1996). Insulin signaling in mice expressing reduced levels of Syp. *J. Biol. Chem.* **271,** 21353–21358.

Bard-Chapeau, E. A., Hevener, A. L., Long, S., Zhang, E. E., Olefsky, J. M., and Feng, G. S. (2005). Deletion of Gab1 in the liver leads to enhanced glucose tolerance and improved hepatic insulin action. *Nat. Med.* **11,** 567–571.

Bard-Chapeau, E. A., Yuan, J., Droin, N., Long, S., Zhang, E. E., Nguyen, T. V., and Feng, G. S. (2006). Concerted functions of Gab1 and Shp2 in liver regeneration and hepatoprotection. *Mol. Cell. Biol.* **26,** 4664–4674.

Barford, D., and Neel, B. G. (1998). Revealing mechanisms for SH2 domain mediated regulation of the protein tyrosine phosphatase SHP-2. *Structure* **6,** 249–254.

Bates, S. H., Stearns, W. H., Dundon, T. A., Schubert, M., Tso, A. W., Wang, Y., Banks, A. S., Lavery, H. J., Haq, A. K., Maratos-Flier, E., Neel, B. G., Schwartz, M. W., *et al.* (2003). STAT3 signalling is required for leptin regulation of energy balance but not reproduction. *Nature* **421,** 856–859.

Bentires-Alj, M., Paez, J. G., David, F. S., Keilhack, H., Halmos, B., Naoki, K., Maris, J. M., Richardson, A., Bardelli, A., Sugarbaker, D. J., Richards, W. G., Du, J., *et al.* (2004). Activating mutations of the Noonan syndrome-associated SHP2/PTPN11 gene in human solid tumors and adult acute myelogenous leukemia. *Cancer Res.* **64,** 8816–8820.

Berchtold, S., Volarevic, S., Moriggl, R., Mercep, M., and Groner, B. (1998). Dominant negative variants of the SHP-2 tyrosine phosphatase inhibit prolactin activation of Jak2 (janus kinase 2) and induction of Stat5 (signal transducer and activator of transcription 5)-dependent transcription. *Mol. Endocrinol.* **12,** 556–567.

Birchmeier, C. (2009). ErbB receptors and the development of the nervous system. *Exp. Cell Res.* **315,** 611–618.

Bonni, A., Sun, Y., Nadal-Vicens, M., Bhatt, A., Frank, D. A., Rozovsky, I., Stahl, N., Yancopoulos, G. D., and Greenberg, M. E. (1997). Regulation of gliogenesis in the central nervous system by the JAK-STAT signaling pathway. *Science* **278,** 477–483.

Borowiak, M., Garratt, A. N., Wustefeld, T., Strehle, M., Trautwein, C., and Birchmeier, C. (2004). Met provides essential signals for liver regeneration. *Proc. Natl. Acad. Sci. USA* **101**, 10608–10613.

Buckingham, M., Meilhac, S., and Zaffran, S. (2005). Building the mammalian heart from two sources of myocardial cells. *Nat. Rev. Genet.* **6**, 826–835.

Carta, C., Pantaleoni, F., Bocchinfuso, G., Stella, L., Vasta, I., Sarkozy, A., Digilio, C., Palleschi, A., Pizzuti, A., Grammatico, P., Zampino, G., Dallapiccola, B., *et al.* (2006). Germline missense mutations affecting KRAS Isoform B are associated with a severe Noonan syndrome phenotype. *Am. J. Hum. Genet.* **79**, 129–135.

Chan, R. J., Leedy, M. B., Munugalavadla, V., Voorhorst, C. S., Li, Y., Yu, M., and Kapur, R. (2005). Human somatic PTPN11 mutations induce hematopoietic-cell hypersensitivity to granulocyte-macrophage colony-stimulating factor. *Blood* **105**, 3737–3742.

Chan, G., Kalaitzidis, D., and Neel, B. G. (2008). The tyrosine phosphatase Shp2 (PTPN11) in cancer. *Cancer Metastasis Rev.* **27**, 179–192.

Chan, G., Kalaitzidis, D., Usenko, T., Kutok, J. L., Yang, W., Mohi, M. G., and Neel, B. G. (2009). Leukemogenic Ptpn11 causes fatal myeloproliferative disorder via cell-autonomous effects on multiple stages of hematopoiesis. *Blood* **113**, 4414–4424.

Chapman, R. S., Lourenco, P. C., Tonner, E., Flint, D. J., Selbert, S., Takeda, K., Akira, S., Clarke, A. R., and Watson, C. J. (1999). Suppression of epithelial apoptosis and delayed mammary gland involution in mice with a conditional knockout of Stat3. *Genes Dev.* **13**, 2604–2616.

Chen, Y., Wen, R., Yang, S., Schuman, J., Zhang, E. E., Yi, T., Feng, G. S., and Wang, D. (2003). Identification of Shp-2 as a Stat5A phosphatase. *J. Biol. Chem.* **278**, 16520–16527.

Choong, K., Freedman, M. H., Chitayat, D., Kelly, E. N., Taylor, G., and Zipursky, A. (1999). Juvenile myelomonocytic leukemia and Noonan syndrome. *J. Pediatr. Hematol. Oncol.* **21**, 523–527.

Christiansen, J. H., Coles, E. G., and Wilkinson, D. G. (2000). Molecular control of neural crest formation, migration and differentiation. *Curr. Opin. Cell Biol.* **12**, 719–724.

Chughtai, N., Schimchowitsch, S., Lebrun, J. J., and Ali, S. (2002). Prolactin induces SHP-2 association with Stat5, nuclear translocation, and binding to the beta-casein gene promoter in mammary cells. *J. Biol. Chem.* **277**, 31107–31114.

Cotton, J. L., and Williams, R. G. (1995). Noonan syndrome and neuroblastoma. *Arch. Pediatr. Adolesc. Med.* **149**, 1280–1281.

Crane, J. F., and Trainor, P. A. (2006). Neural crest stem and progenitor cells. *Annu. Rev. Cell Dev. Biol.* **22**, 267–286.

Cressman, D. E., Greenbaum, L. E., DeAngelis, R. A., Ciliberto, G., Furth, E. E., Poli, V., and Taub, R. (1996). Liver failure and defective hepatocyte regeneration in interleukin-6-deficient mice. *Science* **274**, 1379–1383.

Crompton, T., Gilmour, K. C., and Owen, M. J. (1996). The MAP kinase pathway controls differentiation from double-negative to double-positive thymocyte. *Cell* **86**, 243–251.

Cross, J. C. (2000). Genetic insights into trophoblast differentiation and placental morphogenesis. *Semin. Cell Dev. Biol.* **11**, 105–113.

Digilio, M. C., Pacileo, G., Sarkozy, A., Limongelli, G., Conti, E., Cerrato, F., Marino, B., Pizzuti, A., Calabro, R., and Dallapiccola, B. (2004). Familial aggregation of genetically heterogeneous hypertrophic cardiomyopathy: a boy with LEOPARD syndrome due to PTPN11 mutation and his nonsyndromic father lacking PTPN11 mutations. *Birth Defects Res. A Clin. Mol. Teratol.* **70**, 95–98.

Dong, Z., Brennan, A., Liu, N., Yarden, Y., Lefkowitz, G., Mirsky, R., and Jessen, K. R. (1995). Neu differentiation factor is a neuron-glia signal and regulates survival, proliferation, and maturation of rat Schwann cell precursors. *Neuron* **15**, 585–596.

Emanuel, P. D., Bates, L. J., Castleberry, R. P., Gualtieri, R. J., and Zuckerman, K. S. (1991). Selective hypersensitivity to granulocyte-macrophage colony-stimulating factor by juvenile chronic myeloid leukemia hematopoietic progenitors. *Blood* **77**, 925–929.

Emanuel, P. D., Shannon, K. M., and Castleberry, R. P. (1996). Juvenile myelomonocytic leukemia: molecular understanding and prospects for therapy. *Mol. Med. Today* **2**, 468–475.

Fausto, N., and Campbell, J. S. (2003). The role of hepatocytes and oval cells in liver regeneration and repopulation. *Mech. Dev.* **120**, 117–130.

Fausto, N., Laird, A. D., and Webber, E. M. (1995). Liver regeneration. 2. Role of growth factors and cytokines in hepatic regeneration. *FASEB J.* **9**, 1527–1536.

Feldman, B., Poueymirou, W., Papaioannou, V. E., DeChiara, T. M., and Goldfarb, M. (1995). Requirement of FGF-4 for postimplantation mouse development. *Science* **267**, 246–249.

Feng, G. S. (2007). Shp2-mediated molecular signaling in control of embryonic stem cell self-renewal and differentiation. *Cell Res.* **17**, 37–41.

Feng, G. S., Hui, C. C., and Pawson, T. (1993). SH2-containing phosphotyrosine phosphatase as a target of protein-tyrosine kinases. *Science* **259**, 1607–1611.

Fornaro, M., Burch, P. M., Yang, W., Zhang, L., Hamilton, C. E., Kim, J. H., Neel, B. G., and Bennett, A. M. (2006). SHP-2 activates signaling of the nuclear factor of activated T cells to promote skeletal muscle growth. *J. Cell Biol.* **175**, 87–97.

Fragale, A., Tartaglia, M., Wu, J., and Gelb, B. D. (2004). Noonan syndrome-associated SHP2/PTPN11 mutants cause EGF-dependent prolonged GAB1 binding and sustained ERK2/MAPK1 activation. *Hum. Mutat.* **23**, 267–277.

Freeman, R. M., Jr., Plutzky, J., and Neel, B. G. (1992). Identification of a human src homology 2-containing protein-tyrosine-phosphatase: a putative homolog of Drosophila corkscrew. *Proc. Natl. Acad. Sci. USA* **89**, 11239–11243.

Gao, Q., Wolfgang, M. J., Neschen, S., Morino, K., Horvath, T. L., Shulman, G. I., and Fu, X. Y. (2004). Disruption of neural signal transducer and activator of transcription 3 causes obesity, diabetes, infertility, and thermal dysregulation. *Proc. Natl. Acad. Sci. USA* **101**, 4661–4666.

Garratt, A. N., Voiculescu, O., Topilko, P., Charnay, P., and Birchmeier, C. (2000). A dual role of erbB2 in myelination and in expansion of the schwann cell precursor pool. *J. Cell Biol.* **148**, 1035–1046.

Gauthier, A. S., Furstoss, O., Araki, T., Chan, R., Neel, B. G., Kaplan, D. R., and Miller, F. D. (2007). Control of CNS cell-fate decisions by SHP-2 and its dysregulation in Noonan syndrome. *Neuron* **54**, 245–262.

Goldin, S. N., and Papaioannou, V. E. (2003). Paracrine action of FGF4 during periimplantation development maintains trophectoderm and primitive endoderm. *Genesis* **36**, 40–47.

Gotoh, N., Manova, K., Tanaka, S., Murohashi, M., Hadari, Y., Lee, A., Hamada, Y., Hiroe, T., Ito, M., Kurihara, T., Nakazato, H., Shibuya, M., et al. (2005). The docking protein FRS2alpha is an essential component of multiple fibroblast growth factor responses during early mouse development. *Mol. Cell. Biol.* **25**, 4105–4116.

Grossmann, K. S., Wende, H., Paul, F. E., Cheret, C., Garratt, A. N., Zurborg, S., Feinberg, K., Besser, D., Schulz, H., Peles, E., Selbach, M., Birchmeier, W., et al. (2009). The tyrosine phosphatase Shp2 (PTPN11) directs Neuregulin-1/ErbB signaling throughout Schwann cell development. *Proc. Natl. Acad. Sci. USA* **106**, 16704–16709.

Gutch, M. J., Flint, A. J., Keller, J., Tonks, N. K., and Hengartner, M. O. (1998). The Caenorhabditis elegans SH2 domain-containing protein tyrosine phosphatase PTP-2 participates in signal transduction during oogenesis and vulval development. *Genes Dev.* **12**, 571–585.

Hadari, Y. R., Kouhara, H., Lax, I., and Schlessinger, J. (1998). Binding of Shp2 tyrosine phosphatase to FRS2 is essential for fibroblast growth factor-induced PC12 cell differentiation. *Mol. Cell. Biol.* **18**, 3966–3973.

Hagihara, K., Zhang, E. E., Ke, Y. H., Liu, G., Liu, J. J., Rao, Y., and Feng, G. S. (2009). Shp2 acts downstream of SDF-1alpha/CXCR4 in guiding granule cell migration during cerebellar development. *Dev. Biol.* **334**, 276–284.

Hanna, N., Montagner, A., Lee, W. H., Miteva, M., Vidal, M., Vidaud, M., Parfait, B., and Raynal, P. (2006). Reduced phosphatase activity of SHP-2 in LEOPARD syndrome: consequences for PI3K binding on Gab1. *FEBS Lett.* **580**, 2477–2482.

Hennighausen, L., and Robinson, G. W. (2005). Information networks in the mammary gland. *Nat. Rev. Mol. Cell Biol.* **6**, 715–725.

Herbst, R., Carroll, P. M., Allard, J. D., Schilling, J., Raabe, T., and Simon, M. A. (1996). Daughter of sevenless is a substrate of the phosphotyrosine phosphatase Corkscrew and functions during sevenless signaling. *Cell* **85**, 899–909.

Hof, P., Pluskey, S., Dhe-Paganon, S., Eck, M. J., and Shoelson, S. E. (1998). Crystal structure of the tyrosine phosphatase SHP-2. *Cell* **92**, 441–450.

Holgado-Madruga, M., Emlet, D. R., Moscatello, D. K., Godwin, A. K., and Wong, A. J. (1996). A Grb2-associated docking protein in EGF- and insulin-receptor signalling. *Nature* **379**, 560–564.

Horsley, V., Friday, B. B., Matteson, S., Kegley, K. M., Gephart, J., and Pavlath, G. K. (2001). Regulation of the growth of multinucleated muscle cells by an NFATC2-dependent pathway. *J. Cell Biol.* **153**, 329–338.

Huh, C. G., Factor, V. M., Sanchez, A., Uchida, K., Conner, E. A., and Thorgeirsson, S. S. (2004). Hepatocyte growth factor/c-met signaling pathway is required for efficient liver regeneration and repair. *Proc. Natl. Acad. Sci. USA* **101**, 4477–4482.

Ijiri, R., Tanaka, Y., Keisuke, K., Masuno, M., and Imaizumi, K. (2000). A case of Noonan's syndrome with possible associated neuroblastoma. *Pediatr. Radiol.* **30**, 432–433.

Ivins Zito, C., Kontaridis, M. I., Fornaro, M., Feng, G. S., and Bennett, A. M. (2004). SHP-2 regulates the phosphatidylinositide 3'-kinase/Akt pathway and suppresses caspase 3-mediated apoptosis. *J. Cell. Physiol.* **199**, 227–236.

Jessen, K. R., and Mirsky, R. (2005). The origin and development of glial cells in peripheral nerves. *Nat. Rev. Neurosci.* **6**, 671–682.

Jiang, X., Rowitch, D. H., Soriano, P., McMahon, A. P., and Sucov, H. M. (2000). Fate of the mammalian cardiac neural crest. *Development* **127**, 1607–1616.

Johe, K. K., Hazel, T. G., Muller, T., Dugich-Djordjevic, M. M., and McKay, R. D. (1996). Single factors direct the differentiation of stem cells from the fetal and adult central nervous system. *Genes Dev.* **10**, 3129–3140.

Ke, Y., Lesperance, J., Zhang, E. E., Bard-Chapeau, E. A., Oshima, R. G., Muller, W. J., and Feng, G. S. (2006). Conditional deletion of Shp2 in the mammary gland leads to impaired lobulo-alveolar outgrowth and attenuated Stat5 activation. *J. Biol. Chem.* **281**, 34374–34380.

Ke, Y., Zhang, E. E., Hagihara, K., Wu, D., Pang, Y., Klein, R., Curran, T., Ranscht, B., and Feng, G. S. (2007). Deletion of Shp2 in the brain leads to defective proliferation and differentiation in neural stem cells and early postnatal lethality. *Mol. Cell. Biol.* **27**, 6706–6717.

Kegley, K. M., Gephart, J., Warren, G. L., and Pavlath, G. K. (2001). Altered primary myogenesis in NFATC3(-/-) mice leads to decreased muscle size in the adult. *Dev. Biol.* **232**, 115–126.

Keilhack, H., David, F. S., McGregor, M., Cantley, L. C., and Neel, B. G. (2005). Diverse biochemical properties of Shp2 mutants. Implications for disease phenotypes. *J. Biol. Chem.* **280**, 30984–30993.

Keren, B., Hadchouel, A., Saba, S., Sznajer, Y., Bonneau, D., Leheup, B., Boute, O., Gaillard, D., Lacombe, D., Layet, V., Marlin, S., Mortier, G., *et al.* (2004). PTPN11 mutations in patients with LEOPARD syndrome: a French multicentric experience. *J. Med. Genet.* **41**, e117.

Kirby, M. L., and Waldo, K. L. (1990). Role of neural crest in congenital heart disease. *Circulation* **82**, 332–340.

Kirby, M. L., Gale, T. F., and Stewart, D. E. (1983). Neural crest cells contribute to normal aorticopulmonary septation. *Science* **220**, 1059–1061.

Kitamura, T., Nakae, J., Kitamura, Y., Kido, Y., Biggs, W. H., 3rd, Wright, C. V., White, M. F., Arden, K. C., and Accili, D. (2002). The forkhead transcription factor Foxo1 links insulin signaling to Pdx1 regulation of pancreatic beta cell growth. *J. Clin. Invest.* **110**, 1839–1847.

Kontaridis, M. I., Eminaga, S., Fornaro, M., Zito, C. I., Sordella, R., Settleman, J., and Bennett, A. M. (2004). SHP-2 positively regulates myogenesis by coupling to the Rho GTPase signaling pathway. *Mol. Cell. Biol.* **24**, 5340–5352.

Kontaridis, M. I., Swanson, K. D., David, F. S., Barford, D., and Neel, B. G. (2006). PTPN11 (Shp2) mutations in LEOPARD syndrome have dominant negative, not activating, effects. *J. Biol. Chem.* **281**, 6785–6792.

Kontaridis, M. I., Yang, W., Bence, K. K., Cullen, D., Wang, B., Bodyak, N., Ke, Q., Hinek, A., Kang, P. M., Liao, R., and Neel, B. G. (2008). Deletion of Ptpn11 (Shp2) in cardiomyocytes causes dilated cardiomyopathy via effects on the extracellular signal-regulated kinase/mitogen-activated protein kinase and RhoA signaling pathways. *Circulation* **117**, 1423–1435.

Krajewska, M., Banares, S., Zhang, E. E., Huang, X., Scadeng, M., Jhala, U. S., Feng, G. S., and Krajewski, S. (2008). Development of diabesity in mice with neuronal deletion of Shp2 tyrosine phosphatase. *Am. J. Pathol.* **172**, 1312–1324.

Krenz, M., Gulick, J., Osinska, H. E., Colbert, M. C., Molkentin, J. D., and Robbins, J. (2008). Role of ERK1/2 signaling in congenital valve malformations in Noonan syndrome. *Proc. Natl. Acad. Sci. USA* **105**, 18930–18935.

Lawrence, M., Shao, C., Duan, L., McGlynn, K., and Cobb, M. H. (2008). The protein kinases ERK1/2 and their roles in pancreatic beta cells. *Acta Physiol. (Oxf.)* **192**, 11–17.

Le Douarin, N. M., Creuzet, S., Couly, G., and Dupin, E. (2004). Neural crest cell plasticity and its limits. *Development* **131**, 4637–4650.

Lemke, G. (2006). Neuregulin-1 and myelination. *Sci. STKE* **2006**, pe11.

Lillien, L. (1998). Progenitor cells: what do they know and when do they know it? *Curr. Biol.* **8**, R872–R874.

Liu, X., Robinson, G. W., Wagner, K. U., Garrett, L., Wynshaw-Boris, A., and Hennighausen, L. (1997). Stat5a is mandatory for adult mammary gland development and lactogenesis. *Genes Dev.* **11**, 179–186.

Loh, M. L., Reynolds, M. G., Vattikuti, S., Gerbing, R. B., Alonzo, T. A., Carlson, E., Cheng, J. W., Lee, C. M., Lange, B. J., and Meshinchi, S. (2004a). PTPN11 mutations in pediatric patients with acute myeloid leukemia: results from the Children's Cancer Group. *Leukemia* **18**, 1831–1834.

Loh, M. L., Vattikuti, S., Schubbert, S., Reynolds, M. G., Carlson, E., Lieuw, K. H., Cheng, J. W., Lee, C. M., Stokoe, D., Bonifas, J. M., Curtiss, N. P., Gotlib, J., *et al.* (2004b). Mutations in PTPN11 implicate the SHP-2 phosphatase in leukemogenesis. *Blood* **103**, 2325–2331.

Lopez-Miranda, B., Westra, S. J., Yazdani, S., and Boechat, M. I. (1997). Noonan syndrome associated with neuroblastoma: a case report. *Pediatr. Radiol.* **27**, 324–326.

Lyons, D. A., Pogoda, H. M., Voas, M. G., Woods, I. G., Diamond, B., Nix, R., Arana, N., Jacobs, J., and Talbot, W. S. (2005). erbb3 and erbb2 are essential for schwann cell migration and myelination in zebrafish. *Curr. Biol.* **15**, 513–524.

Marino, B., Digilio, M. C., Toscano, A., Giannotti, A., and Dallapiccola, B. (1999). Congenital heart diseases in children with Noonan syndrome: An expanded cardiac spectrum with high prevalence of atrioventricular canal. *J. Pediatr.* **135**, 703–706.

Maroun, C. R., Naujokas, M. A., Holgado-Madruga, M., Wong, A. J., and Park, M. (2000). The tyrosine phosphatase SHP-2 is required for sustained activation of extracellular signal-regulated kinase and epithelial morphogenesis downstream from the met receptor tyrosine kinase. *Mol. Cell. Biol.* **20**, 8513–8525.

Martinelli, S., Carta, C., Flex, E., Binni, F., Cordisco, E. L., Moretti, S., Puxeddu, E., Tonacchera, M., Pinchera, A., McDowell, H. P., Dominici, C., Rosolen, A., *et al.* (2006). Activating PTPN11 mutations play a minor role in pediatric and adult solid tumors. *Cancer Genet. Cytogenet.* **166**, 124–129.

Menard, C., Hein, P., Paquin, A., Savelson, A., Yang, X. M., Lederfein, D., Barnabe-Heider, F., Mir, A. A., Sterneck, E., Peterson, A. C., Johnson, P. F., Vinson, C., *et al.* (2002). An essential role for a MEK-C/EBP pathway during growth factor-regulated cortical neurogenesis. *Neuron* **36**, 597–610.

Meyer, D., and Birchmeier, C. (1995). Multiple essential functions of neuregulin in development. *Nature* **378**, 386–390.

Michailov, G. V., Sereda, M. W., Brinkmann, B. G., Fischer, T. M., Haug, B., Birchmeier, C., Role, L., Lai, C., Schwab, M. H., and Nave, K. A. (2004). Axonal neuregulin-1 regulates myelin sheath thickness. *Science* **304**, 700–703.

Milarski, K. L., and Saltiel, A. R. (1994). Expression of catalytically inactive Syp phosphatase in 3T3 cells blocks stimulation of mitogen-activated protein kinase by insulin. *J. Biol. Chem.* **269**, 21239–21243.

Miller, F. D., and Gauthier, A. S. (2007). Timing is everything: making neurons versus glia in the developing cortex. *Neuron* **54**, 357–369.

Miyamoto, D., Miyamoto, M., Takahashi, A., Yomogita, Y., Higashi, H., Kondo, S., and Hatakeyama, M. (2008). Isolation of a distinct class of gain-of-function SHP-2 mutants with oncogenic RAS-like transforming activity from solid tumors. *Oncogene* **27**, 3508–3515.

Mohi, M. G., Williams, I. R., Dearolf, C. R., Chan, G., Kutok, J. L., Cohen, S., Morgan, K., Boulton, C., Shigematsu, H., Keilhack, H., Akashi, K., Gilliland, D. G., *et al.* (2005). Prognostic, therapeutic, and mechanistic implications of a mouse model of leukemia evoked by Shp2 (PTPN11) mutations. *Cancer Cell* **7**, 179–191.

Molina, T. J., Kishihara, K., Siderovski, D. P., van Ewijk, W., Narendran, A., Timms, E., Wakeham, A., Paige, C. J., Hartmann, K. U., Veillette, A., *et al.* (1992). Profound block in thymocyte development in mice lacking p56lck. *Nature* **357**, 161–164.

Montagner, A., Yart, A., Dance, M., Perret, B., Salles, J. P., and Raynal, P. (2005). A novel role for Gab1 and SHP2 in epidermal growth factor-induced Ras activation. *J. Biol. Chem.* **280**, 5350–5360.

Morris, J. K., Lin, W., Hauser, C., Marchuk, Y., Getman, D., and Lee, K. F. (1999). Rescue of the cardiac defect in ErbB2 mutant mice reveals essential roles of ErbB2 in peripheral nervous system development. *Neuron* **23**, 273–283.

Musante, L., Kehl, H. G., Majewski, F., Meinecke, P., Schweiger, S., Gillessen-Kaesbach, G., Wieczorek, D., Hinkel, G. K., Tinschert, S., Hoeltzenbein, M., Ropers, H. H., and Kalscheuer, V. M. (2003). Spectrum of mutations in PTPN11 and genotype-phenotype correlation in 96 patients with Noonan syndrome and five patients with cardio-facio-cutaneous syndrome. *Eur. J. Hum. Genet.* **11**, 201–206.

Nakae, J., Biggs, W. H., 3rd, Kitamura, T., Cavenee, W. K., Wright, C. V., Arden, K. C., and Accili, D. (2002). Regulation of insulin action and pancreatic beta-cell function by mutated alleles of the gene encoding forkhead transcription factor Foxo1. *Nat. Genet.* **32**, 245–253.

Nakamura, T., Colbert, M., Krenz, M., Molkentin, J. D., Hahn, H. S., Dorn, G. W., 2nd, and Robbins, J. (2007). Mediating ERK 1/2 signaling rescues congenital heart defects in a mouse model of Noonan syndrome. *J. Clin. Invest.* **117**, 2123–2132.

Nakamura, T., Gulick, J., Colbert, M. C., and Robbins, J. (2009a). Protein tyrosine phosphatase activity in the neural crest is essential for normal heart and skull development. *Proc. Natl. Acad. Sci. USA* **106,** 11270–11275.

Nakamura, T., Gulick, J., Pratt, R., and Robbins, J. (2009b). Noonan syndrome is associated with enhanced pERK activity, the repression of which can prevent craniofacial malformations. *Proc. Natl. Acad. Sci. USA* **106,** 15436–15441.

Nakashima, K., Yanagisawa, M., Arakawa, H., Kimura, N., Hisatsune, T., Kawabata, M., Miyazono, K., and Taga, T. (1999). Synergistic signaling in fetal brain by STAT3-Smad1 complex bridged by p300. *Science* **284,** 479–482.

Neel, B. G., Gu, H., and Pao, L. (2003). The 'Shp'ing news: SH2 domain-containing tyrosine phosphatases in cell signaling. *Trends Biochem. Sci.* **28,** 284–293.

Newbern, J., Zhong, J., Wickramasinghe, R. S., Li, X., Wu, Y., Samuels, I., Cherosky, N., Karlo, J. C., O'Loughlin, B., Wikenheiser, J., Gargesha, M., Doughman, Y. Q., et al. (2008). Mouse and human phenotypes indicate a critical conserved role for ERK2 signaling in neural crest development. *Proc. Natl. Acad. Sci. USA* **105,** 17115–17120.

Nguyen, T. V., Ke, Y., Zhang, E. E., and Feng, G. S. (2006). Conditional deletion of Shp2 tyrosine phosphatase in thymocytes suppresses both pre-TCR and TCR signals. *J. Immunol.* **177,** 5990–5996.

Niihori, T., Aoki, Y., Ohashi, H., Kurosawa, K., Kondoh, T., Ishikiriyama, S., Kawame, H., Kamasaki, H., Yamanaka, T., Takada, F., Nishio, K., Sakurai, M., et al. (2005). Functional analysis of PTPN11/SHP-2 mutants identified in Noonan syndrome and childhood leukemia. *J. Hum. Genet.* **50,** 192–202.

Noguchi, T., Matozaki, T., Horita, K., Fujioka, Y., and Kasuga, M. (1994). Role of SH-PTP2, a protein-tyrosine phosphatase with Src homology 2 domains, in insulin-stimulated Ras activation. *Mol. Cell. Biol.* **14,** 6674–6682.

Noonan, J. A. (1968). Hypertelorism with Turner phenotype. A new syndrome with associated congenital heart disease. *Am. J. Dis. Child* **116,** 373–380.

Noonan, J. A. (2006). Noonan syndrome and related disorders: alterations in growth and puberty. *Rev. Endocr. Metab. Disord.* **7,** 251–255.

Noonan, J., and Ehmke, D. (1963). Associated non cardiac malformations in children with congenital heart disease. *J. Pediatr.* **63,** 468–470.

Ogata, T., and Yoshida, R. (2005). PTPN11 mutations and genotype-phenotype correlations in Noonan and LEOPARD syndromes. *Pediatr. Endocrinol. Rev.* **2,** 669–674.

Oh, E. S., Gu, H., Saxton, T. M., Timms, J. F., Hausdorff, S., Frevert, E. U., Kahn, B. B., Pawson, T., Neel, B. G., and Thomas, S. M. (1999). Regulation of early events in integrin signaling by protein tyrosine phosphatase SHP-2. *Mol. Cell. Biol.* **19,** 3205–3215.

Oishi, K., Gaengel, K., Krishnamoorthy, S., Kamiya, K., Kim, I. K., Ying, H., Weber, U., Perkins, L. A., Tartaglia, M., Mlodzik, M., Pick, L., and Gelb, B. D. (2006). Transgenic Drosophila models of Noonan syndrome causing PTPN11 gain-of-function mutations. *Hum. Mol. Genet.* **15,** 543–553.

Oishi, K., Zhang, H., Gault, W. J., Wang, C. J., Tan, C. C., Kim, I. K., Ying, H., Rahman, T., Pica, N., Tartaglia, M., Mlodzik, M., and Gelb, B. D. (2009). Phosphatase-defective LEOPARD syndrome mutations in PTPN11 gene have gain-of-function effects during Drosophila development. *Hum. Mol. Genet.* **18,** 193–201.

Opitz, J. M. (1985). The Noonan syndrome. *Am. J. Med. Genet.* **21,** 515–518.

O'Reilly, A. M., Pluskey, S., Shoelson, S. E., and Neel, B. G. (2000). Activated mutants of SHP-2 preferentially induce elongation of Xenopus animal caps. *Mol. Cell. Biol.* **20,** 299–311.

Pagani, M. R., Oishi, K., Gelb, B. D., and Zhong, Y. (2009). The phosphatase SHP2 regulates the spacing effect for long-term memory induction. *Cell* **139,** 186–198.

Pandit, B., Sarkozy, A., Pennacchio, L. A., Carta, C., Oishi, K., Martinelli, S., Pogna, E. A., Schackwitz, W., Ustaszewska, A., Landstrom, A., Bos, J. M., Ommen, S. R., *et al.* (2007). Gain-of-function RAF1 mutations cause Noonan and LEOPARD syndromes with hypertrophic cardiomyopathy. *Nat. Genet.* **39**, 1007–1012.

Park, J. K., Williams, B. P., Alberta, J. A., and Stiles, C. D. (1999). Bipotent cortical progenitor cells process conflicting cues for neurons and glia in a hierarchical manner. *J. Neurosci.* **19**, 10383–10389.

Perkins, L. A., Larsen, I., and Perrimon, N. (1992). Corkscrew encodes a putative protein tyrosine phosphatase that functions to transduce the terminal signal from the receptor tyrosine kinase torso. *Cell* **70**, 225–236.

Perkins, L. A., Johnson, M. R., Melnick, M. B., and Perrimon, N. (1996). The nonreceptor protein tyrosine phosphatase corkscrew functions in multiple receptor tyrosine kinase pathways in Drosophila. *Dev. Biol.* **180**, 63–81.

Princen, F., Bard, E., Sheikh, F., Zhang, S. S., Wang, J., Zago, W. M., Wu, D., Trelles, R. D., Bailly-Maitre, B., Kahn, C. R., Chen, Y., Reed, J. C., *et al.* (2009). Deletion of Shp2 tyrosine phosphatase in muscle leads to dilated cardiomyopathy, insulin resistance, and premature death. *Mol. Cell. Biol.* **29**, 378–388.

Qu, C. K. (2002). Role of the SHP-2 tyrosine phosphatase in cytokine-induced signaling and cellular response. *Biochim. Biophys. Acta* **1592**, 297–301.

Qu, C. K., Shi, Z. Q., Shen, R., Tsai, F. Y., Orkin, S. H., and Feng, G. S. (1997). A deletion mutation in the SH2-N domain of Shp-2 severely suppresses hematopoietic cell development. *Mol. Cell. Biol.* **17**, 5499–5507.

Qu, C. K., Yu, W. M., Azzarelli, B., Cooper, S., Broxmeyer, H. E., and Feng, G. S. (1998). Biased suppression of hematopoiesis and multiple developmental defects in chimeric mice containing Shp-2 mutant cells. *Mol. Cell. Biol.* **18**, 6075–6082.

Qu, C. K., Nguyen, S., Chen, J., and Feng, G. S. (2001). Requirement of Shp-2 tyrosine phosphatase in lymphoid and hematopoietic cell development. *Blood* **97**, 911–914.

Raballo, R., Rhee, J., Lyn-Cook, R., Leckman, J. F., Schwartz, M. L., and Vaccarino, F. M. (2000). Basic fibroblast growth factor (Fgf2) is necessary for cell proliferation and neurogenesis in the developing cerebral cortex. *J. Neurosci.* **20**, 5012–5023.

Riethmacher, D., Sonnenberg-Riethmacher, E., Brinkmann, V., Yamaai, T., Lewin, G. R., and Birchmeier, C. (1997). Severe neuropathies in mice with targeted mutations in the ErbB3 receptor. *Nature* **389**, 725–730.

Roberts, A. E., Araki, T., Swanson, K. D., Montgomery, K. T., Schiripo, T. A., Joshi, V. A., Li, L., Yassin, Y., Tamburino, A. M., Neel, B. G., and Kucherlapati, R. S. (2007). Germline gain-of-function mutations in SOS1 cause Noonan syndrome. *Nat. Genet.* **39**, 70–74.

Rosário, M., and Birchmeier, W. (2003). How to make tubes: signaling by the Met receptor tyrosine kinase. *Trends Cell Biol.* **13**, 328–335.

Rosário, M., Franke, R., Bednarski, C., and Birchmeier, W. (2007). The neurite outgrowth multiadaptor RhoGAP, NOMA-GAP, regulates neurite extension through SHP2 and Cdc42. *J. Cell Biol.* **178**, 503–516.

Rossant, J. (2001). Stem cells from the Mammalian blastocyst. *Stem Cells* **19**, 477–482.

Rossant, J., and Cross, J. C. (2001). Placental development: lessons from mouse mutants. *Nat. Rev. Genet.* **2**, 538–548.

Saba-El-Leil, M. K., Vella, F. D., Vernay, B., Voisin, L., Chen, L., Labrecque, N., Ang, S. L., and Meloche, S. (2003). An essential function of the mitogen-activated protein kinase Erk2 in mouse trophoblast development. *EMBO Rep.* **4**, 964–968.

Saxton, T. M., Henkemeyer, M., Gasca, S., Shen, R., Rossi, D. J., Shalaby, F., Feng, G. S., and Pawson, T. (1997). Abnormal mesoderm patterning in mouse embryos mutant for the SH2 tyrosine phosphatase Shp-2. *EMBO J.* **16**, 2352–2364.

Schaeper, U., Gehring, N. H., Fuchs, K. P., Sachs, M., Kempkes, B., and Birchmeier, W. (2000). Coupling of Gab1 to c-Met, Grb2, and Shp2 mediates biological responses. *J. Cell Biol.* **149**, 1419–1432.

Schoenwaelder, S. M., Petch, L. A., Williamson, D., Shen, R., Feng, G. S., and Burridge, K. (2000). The protein tyrosine phosphatase Shp-2 regulates RhoA activity. *Curr. Biol.* **10**, 1523–1526.

Schubbert, S., Lieuw, K., Rowe, S. L., Lee, C. M., Li, X., Loh, M. L., Clapp, D. W., and Shannon, K. M. (2005). Functional analysis of leukemia-associated PTPN11 mutations in primary hematopoietic cells. *Blood* **106**, 311–317.

Schubbert, S., Zenker, M., Rowe, S. L., Boll, S., Klein, C., Bollag, G., van der Burgt, I., Musante, L., Kalscheuer, V., Wehner, L. E., Nguyen, H., West, B., *et al.* (2006). Germline KRAS mutations cause Noonan syndrome. *Nat. Genet.* **38**, 331–336.

Shen, Q., Qian, X., Capela, A., and Temple, S. (1998). Stem cells in the embryonic cerebral cortex: their role in histogenesis and patterning. *J. Neurobiol.* **36**, 162–174.

Shi, Z. Q., Lu, W., and Feng, G. S. (1998). The Shp-2 tyrosine phosphatase has opposite effects in mediating the activation of extracellular signal-regulated and c-Jun NH2-terminal mitogen-activated protein kinases. *J. Biol. Chem.* **273**, 4904–4908.

Srivastava, D. (2006). Making or breaking the heart: from lineage determination to morphogenesis. *Cell* **126**, 1037–1048.

Sun, Y., Nadal-Vicens, M., Misono, S., Lin, M. Z., Zubiaga, A., Hua, X., Fan, G., and Greenberg, M. E. (2001). Neurogenin promotes neurogenesis and inhibits glial differentiation by independent mechanisms. *Cell* **104**, 365–376.

Swat, W., Shinkai, Y., Cheng, H. L., Davidson, L., and Alt, F. W. (1996). Activated Ras signals differentiation and expansion of $CD4^+8^+$ thymocytes. *Proc. Natl. Acad. Sci. USA* **93**, 4683–4687.

Tanaka, S., Kunath, T., Hadjantonakis, A. K., Nagy, A., and Rossant, J. (1998). Promotion of trophoblast stem cell proliferation by FGF4. *Science* **282**, 2072–2075.

Tang, T. L., Freeman, R. M., Jr., O'Reilly, A. M., Neel, B. G., and Sokol, S. Y. (1995). The SH2-containing protein-tyrosine phosphatase SH-PTP2 is required upstream of MAP kinase for early Xenopus development. *Cell* **80**, 473–483.

Tartaglia, M., and Gelb, B. D. (2005). Noonan syndrome and related disorders: genetics and pathogenesis. *Annu. Rev. Genomics Hum. Genet.* **6**, 45–68.

Tartaglia, M., Mehler, E. L., Goldberg, R., Zampino, G., Brunner, H. G., Kremer, H., van der Burgt, I., Crosby, A. H., Ion, A., Jeffery, S., Kalidas, K., Patton, M. A., *et al.* (2001). Mutations in PTPN11, encoding the protein tyrosine phosphatase SHP-2, cause Noonan syndrome. *Nat. Genet.* **29**, 465–468.

Tartaglia, M., Kalidas, K., Shaw, A., Song, X., Musat, D. L., van der Burgt, I., Brunner, H. G., Bertola, D. R., Crosby, A., Ion, A., Kucherlapati, R. S., Jeffery, S., *et al.* (2002). PTPN11 mutations in Noonan syndrome: molecular spectrum, genotype-phenotype correlation, and phenotypic heterogeneity. *Am. J. Hum. Genet.* **70**, 1555–1563.

Tartaglia, M., Niemeyer, C. M., Fragale, A., Song, X., Buechner, J., Jung, A., Hahlen, K., Hasle, H., Licht, J. D., and Gelb, B. D. (2003). Somatic mutations in PTPN11 in juvenile myelomonocytic leukemia, myelodysplastic syndromes and acute myeloid leukemia. *Nat. Genet.* **34**, 148–150.

Tartaglia, M., Martinelli, S., Cazzaniga, G., Cordeddu, V., Iavarone, I., Spinelli, M., Palmi, C., Carta, C., Pession, A., Arico, M., Masera, G., Basso, G., *et al.* (2004). Genetic evidence for lineage-related and differentiation stage-related contribution of somatic PTPN11 mutations to leukemogenesis in childhood acute leukemia. *Blood* **104**, 307–313.

Tartaglia, M., Martinelli, S., Iavarone, I., Cazzaniga, G., Spinelli, M., Giarin, E., Petrangeli, V., Carta, C., Masetti, R., Arico, M., Locatelli, F., Basso, G., *et al.* (2005). Somatic PTPN11 mutations in childhood acute myeloid leukaemia. *Br. J. Haematol.* **129**, 333–339.

Tartaglia, M., Martinelli, S., Stella, L., Bocchinfuso, G., Flex, E., Cordeddu, V., Zampino, G., Burgt, I., Palleschi, A., Petrucci, T. C., Sorcini, M., Schoch, C., et al. (2006). Diversity and functional consequences of germline and somatic PTPN11 mutations in human disease. *Am. J. Hum. Genet.* **78,** 279–290.

Tartaglia, M., Pennacchio, L. A., Zhao, C., Yadav, K. K., Fodale, V., Sarkozy, A., Pandit, B., Oishi, K., Martinelli, S., Schackwitz, W., Ustaszewska, A., Martin, J., et al. (2007). Gain-of-function SOS1 mutations cause a distinctive form of Noonan syndrome. *Nat. Genet.* **39,** 75–79.

Taveggia, C., Zanazzi, G., Petrylak, A., Yano, H., Rosenbluth, J., Einheber, S., Xu, X., Esper, R. M., Loeb, J. A., Shrager, P., Chao, M. V., Falls, D. L., et al. (2005). Neuregulin-1 type III determines the ensheathment fate of axons. *Neuron* **47,** 681–694.

Ueki, K., Okada, T., Hu, J., Liew, C. W., Assmann, A., Dahlgren, G. M., Peters, J. L., Shackman, J. G., Zhang, M., Artner, I., Satin, L. S., Stein, R., et al. (2006). Total insulin and IGF-I resistance in pancreatic beta cells causes overt diabetes. *Nat. Genet.* **38,** 583–588.

Uhlen, P., Burch, P. M., Zito, C. I., Estrada, M., Ehrlich, B. E., and Bennett, A. M. (2006). Gain-of-function/Noonan syndrome SHP-2/Ptpn11 mutants enhance calcium oscillations and impair NFAT signaling. *Proc. Natl. Acad. Sci. USA* **103,** 2160–2165.

Vogel, W., Lammers, R., Huang, J., and Ullrich, A. (1993). Activation of a phosphotyrosine phosphatase by tyrosine phosphorylation. *Science* **259,** 1611–1614.

Watson, C. J., and Khaled, W. T. (2008). Mammary development in the embryo and adult: a journey of morphogenesis and commitment. *Development* **135,** 995–1003.

Williams, B. P., Park, J. K., Alberta, J. A., Muhlebach, S. G., Hwang, G. Y., Roberts, T. M., and Stiles, C. D. (1997). A PDGF-regulated immediate early gene response initiates neuronal differentiation in ventricular zone progenitor cells. *Neuron* **18,** 553–562.

Woldeyesus, M. T., Britsch, S., Riethmacher, D., Xu, L., Sonnenberg-Riethmacher, E., Abou-Rebyeh, F., Harvey, R., Caroni, P., and Birchmeier, C. (1999). Peripheral nervous system defects in erbB2 mutants following genetic rescue of heart development. *Genes Dev.* **13,** 2538–2548.

Wu, T. R., Hong, Y. K., Wang, X. D., Ling, M. Y., Dragoi, A. M., Chung, A. S., Campbell, A. G., Han, Z. Y., Feng, G. S., and Chin, Y. E. (2002). SHP-2 is a dual-specificity phosphatase involved in Stat1 dephosphorylation at both tyrosine and serine residues in nuclei. *J. Biol. Chem.* **277,** 47572–47580.

Wu, D., Pang, Y., Ke, Y., Yu, J., He, Z., Tautz, L., Mustelin, T., Ding, S., Huang, Z., and Feng, G. S. (2009). A conserved mechanism for control of human and mouse embryonic stem cell pluripotency and differentiation by shp2 tyrosine phosphatase. *PLoS ONE* **4,** e4914.

Wuestefeld, T., Klein, C., Streetz, K. L., Betz, U., Lauber, J., Buer, J., Manns, M. P., Muller, W., and Trautwein, C. (2003). Interleukin-6/glycoprotein 130-dependent pathways are protective during liver regeneration. *J. Biol. Chem.* **278,** 11281–11288.

Yamamoto, S., Yoshino, I., Shimazaki, T., Murohashi, M., Hevner, R. F., Lax, I., Okano, H., Shibuya, M., Schlessinger, J., and Gotoh, N. (2005). Essential role of Shp2-binding sites on FRS2alpha for corticogenesis and for FGF2-dependent proliferation of neural progenitor cells. *Proc. Natl. Acad. Sci. USA* **102,** 15983–15988.

Yamamoto, T., Isomura, M., Xu, Y., Liang, J., Yagasaki, H., Kamachi, Y., Kudo, K., Kiyoi, H., Naoe, T., and Kojma, S. (2006). PTPN11, RAS and FLT3 mutations in childhood acute lymphoblastic leukemia. *Leuk. Res.* **30,** 1085–1089.

Yang, W., Klaman, L. D., Chen, B., Araki, T., Harada, H., Thomas, S. M., George, E. L., and Neel, B. G. (2006). An Shp2/SFK/Ras/Erk signaling pathway controls trophoblast stem cell survival. *Dev. Cell* **10,** 317–327.

Yang, Z., Li, Y., Yin, F., and Chan, R. J. (2008). Activating PTPN11 mutants promote hematopoietic progenitor cell-cycle progression and survival. *Exp. Hematol.* **36,** 1285–1296.

Yoshida, R., Hasegawa, T., Hasegawa, Y., Nagai, T., Kinoshita, E., Tanaka, Y., Kanegane, H., Ohyama, K., Onishi, T., Hanew, K., Okuyama, T., Horikawa, R., *et al.* (2004). Protein-tyrosine phosphatase, nonreceptor type 11 mutation analysis and clinical assessment in 45 patients with Noonan syndrome. *J. Clin. Endocrinol. Metab.* **89,** 3359–3364.

You, M., Flick, L. M., Yu, D., and Feng, G. S. (2001). Modulation of the nuclear factor kappa B pathway by Shp-2 tyrosine phosphatase in mediating the induction of interleukin (IL)-6 by IL-1 or tumor necrosis factor. *J. Exp. Med.* **193,** 101–110.

Yu, D. H., Qu, C. K., Henegariu, O., Lu, X., and Feng, G. S. (1998). Protein-tyrosine phosphatase Shp-2 regulates cell spreading, migration, and focal adhesion. *J. Biol. Chem.* **273,** 21125–21131.

Yu, W. M., Daino, H., Chen, J., Bunting, K. D., and Qu, C. K. (2006). Effects of a leukemia-associated gain-of-function mutation of SHP-2 phosphatase on interleukin-3 signaling. *J. Biol. Chem.* **281,** 5426–5434.

Zenker, M., Buheitel, G., Rauch, R., Koenig, R., Bosse, K., Kress, W., Tietze, H. U., Doerr, H. G., Hofbeck, M., Singer, H., Reis, A., and Rauch, A. (2004). Genotype-phenotype correlations in Noonan syndrome. *J. Pediatr.* **144,** 368–374.

Zhang, S. Q., Tsiaras, W. G., Araki, T., Wen, G., Minichiello, L., Klein, R., and Neel, B. G. (2002). Receptor-specific regulation of phosphatidylinositol 3'-kinase activation by the protein tyrosine phosphatase Shp2. *Mol. Cell. Biol.* **22,** 4062–4072.

Zhang, E. E., Chapeau, E., Hagihara, K., and Feng, G. S. (2004). Neuronal Shp2 tyrosine phosphatase controls energy balance and metabolism. *Proc. Natl. Acad. Sci. USA* **101,** 16064–16069.

Zhang, S. S., Hao, E., Yu, J., Liu, W., Wang, J., Levine, F., and Feng, G. S. (2009a). Coordinated regulation by Shp2 tyrosine phosphatase of signaling events controlling insulin biosynthesis in pancreatic beta-cells. *Proc. Natl. Acad. Sci. USA* **106,** 7531–7536.

Zhang, W., Chan, R. J., Chen, H., Yang, Z., He, Y., Zhang, X., Luo, Y., Yin, F., Moh, A., Miller, L. C., Payne, R. M., Zhang, Z. Y., *et al.* (2009b). Negative regulation of Stat3 by activating PTPN11 mutants contributes to the pathogenesis of Noonan syndrome and juvenile myelomonocytic leukemia. *J. Biol. Chem.* **284,** 22353–22363.

CXC Chemokines in Cancer Angiogenesis and Metastases

Ellen C. Keeley,* Borna Mehrad,[†] and Robert M. Strieter[†]

*Division of Cardiology, Department of Medicine
University of Virginia, Charlottesville, Virginia, USA
[†]Division of Pulmonary and Critical Care Medicine, Department of Medicine
University of Virginia, Charlottesville, Virginia, USA

I. Introduction
II. Angiogenic CXC Chemokines and Receptors
III. Angiostatic CXC Chemokines and Receptors
IV. Immunoangiostasis
V. Chemokine-Induced Angiogenesis in Tumor Models
  A. Melanoma
  B. Pancreatic Cancer
  C. Ovarian Cancer
  D. Gastrointestinal Cancer
  E. Bronchogenic Cancer
  F. Prostate Cancer
  G. Glioblastoma
  H. Head and Neck Cancer
  I. Renal Cell Cancer
VI. Chemokines Affect on Cancer Metastases
  A. The CXCL12–CXCR4 Axis in Mediating Homing of Metastases
  B. CXCR7, A Novel Receptor for CXCL11 and CXCL12
VII. Conclusion
  References

The tumor microenvironment is extremely complex that depends on tumor cell interaction with the responding host cells. Angiogenesis, or new blood vessel growth from preexisting vasculature, is a preeminent feature of successful tumor growth of all solid tumors. While a number of factors produced by both the tumor cells and host responding cells have been discovered that regulate angiogenesis, increasing evidence is growing to support the important role of CXC chemokines in this process. As a family of cytokines, the CXC chemokines are pleiotropic in their ability to regulate tumor-associated angiogenesis, as well as cancer cell metastases. In this chapter, we will discuss the disparate activity that CXC chemokines play in regulating cancer-associated angiogenesis and metastases. © 2010 Elsevier Inc.

I. INTRODUCTION

Angiogenesis plays a critical role in the development, growth, and metastatic potential of cancer. Chemokines are a family of small heparin-binding proteins (8–10 kDa in size) that were originally described for their role in mediating leukocyte recruitment to sites of inflammation (Charo and Ransohoff, 2006). Within the chemokine family, there are four subgroups (CXC, CC, CX_3C, and C chemokines) that are defined by the positioning of the conserved cysteines near the amino-terminus. The CXC subgroup, the focus of this review, has been shown to play a pivotal role in angiogenesis in both physiologic and pathologic settings (Keeley *et al.*, 2008; Mehrad *et al.*, 2007; Vandercappellen *et al.*, 2008). The CXC chemokine family is further categorized on the basis of the presence or absence of a three amino acid sequence, glutamic acid-leucine-arginine (called the "ELR" motif) proximal to the CXC sequence. The "ELR" motif is important since the ELR containing (ELR+) CXC chemokines are potent promoters of angiogenesis, while the interferon (IFN)-inducible, non-ELR containing (ELR−) CXC chemokines are potent inhibitors of angiogenesis (Table I) (Strieter *et al.*, 1995). In this review, we will discuss the unique role that the ELR+ and ELR− CXC chemokines play in cancer angiogenesis and metastases.

Table I CXC Chemokine Ligands and Receptors Involved in Cancer Angiogenesis and Metastasis

Systemic name	Human ligand	Receptor	Promotes angiogenesis	Promotes metastasis
ELR+ CXC chemokines				
CXCL1	Gro-α	CXCR2	+	+
CXCL2	Gro-β	CXCR2	+	+
CXCL3	Gro-γ	CXCR2	+	+
CXCL5	ENA-78	CXCR2	+	+
CXCL6	GCP-2	CXCR1/CXCR2	+	+
CXCL7	NAP-2	CXCR2	+	?
CXCL8	IL-8	CXCR1/CXCR2	+	+
ELR− CXC chemokines				
CXCL4	PF-4	CXCR3	−	?
CXCL4L1	PF-4var	?	−	−
CXCL9	Mig	CXCR3	−	?
CXCL10	IP-10	CXCR3	−	−
CXCL11	I-TAC	CXCR3/CXCR7	−	?
CXCL12	SDF-1	CXCR4/CXCR7	?	+
CXCL14	BRAK	?	−	?

II. ANGIOGENIC CXC CHEMOKINES AND RECEPTORS

The angiogenic CXC chemokine family includes CXCL1, CXCL2, CXCL3, CXCL5, CXCL6, CXCL7, and CXCL8 (Table I). In the mouse, all ELR+ CXC chemokines signal via CXCR2, whereas in humans, ELR+ CXC chemokine ligands can signal via both CXCR2 and CXCR1 (Mehrad *et al.*, 2007). CXCR2, however, is considered the major angiogenic receptor in humans since the expression of CXCR2 alone is required for endothelial cell chemotaxis despite the fact that both CXCR1 and CXCR2 are detected on endothelial cells (Addison *et al.*, 2000; Murdoch *et al.*, 1999); and immunoneutralization of CXCR2 blocks the response of human endothelial cells to CXCL8 (Heidemann *et al.*, 2003). Lastly, while only CXCL8 and CXCL6 bind to CXCR1, all the human ELR+ CXC chemokines mediate angiogenesis (Mehrad *et al.*, 2007).

In addition to CXCR2, a unique promiscuous, nonsignaling chemokine receptor, the red blood cell Duffy antigen for chemokines (DARC) binds CXCL1, CXCL5, and CXCL8 and is thought to function as a "decoy" for excess ELR+ CXC angiogenic chemokines, thus creating a less angiogenic environment leading to inhibition of tumor growth and metastasis (Addison *et al.*, 2004). When transfected and overexpressed in a human non-small cell cancer tumor line, and implanted into animals, the DARC-expressing tumors had greater necrosis, decreased blood vessel density, and decreased potential of metastases (Addison *et al.*, 2004); similar findings have been shown using breast cancer cell lines (Wang *et al.*, 2006a). In a transgenic adenocarcinoma mouse model of prostate cancer, DARC-knockout mice developed larger, more aggressive tumors, and the tumors had increased blood vessel density and increased levels of angiogenic ELR+ CXC chemokines compared to wild-type mice (Shen *et al.*, 2006). Moreover, in a separate study, transgenic expression of DARC by mouse endothelial cells resulted in an attenuated angiogenic response to ELR+ CXC chemokines *in vivo* (Du *et al.*, 2002). From a clinical perspective, since approximately 80% of individuals of African descent lack DARC, it has been suggested that the decreased clearance of angiogenic chemokines may be the mechanism behind their increased mortality from prostate cancer (Shen *et al.*, 2006).

III. ANGIOSTATIC CXC CHEMOKINES AND RECEPTORS

The angiostatic CXC chemokine family includes the IFN-inducible CXCL4, CXCL9, CXCL10, CXCL11, and CXCL14 (Table I). The major angiostatic receptor for CXCL9, CXCL10, and CXCL11 is a unique

G protein-coupled receptor called CXCR3 (Ehlert *et al.*, 2004; Loetscher *et al.*, 1998; Luster *et al.*, 1998; Rollins, 1997). More recently, however, two novel receptors derived from alternative splicing of the CXCR3 gene product have been described. In humans, therefore, the CXCR3 receptor exists in at least three distinct mRNA splice variants: CXCR3A, CXCR3B, and CXCR3-alt (Lasagni *et al.*, 2003; Mehrad *et al.*, 2007). In one study, human microvascular endothelial cell line-1, transfected with CXCR3A or CXCR3B, was found to bind to CXCL9, CXCL10, and CXCL11; CXCL4, however, showed high affinity for CXCR3B alone (Lasagni *et al.*, 2003). These investigators concluded, therefore, that CXCR3B acts as a functional receptor for CXCL4. CXCR3-alt, which is generated by posttranscriptional exon skipping, has been shown to exhibit a reduced response to CXCL9 and CXCL10 while maintaining its signaling activity to CXCL11 (Ehlert *et al.*, 2004). The roles of the receptors CXCR3-alt and CXCR7 (a novel receptor for CXCL11 and CXCL12) (Balabanian *et al.*, 2005; Burns *et al.*, 2006) in regard to angiogenesis remain undefined.

CXCR4 is the main receptor for CXCL12, a non-ELR containing CXC chemokine. The CXCL12–CXCR4 biological axis has been associated with tumor invasion and metastases (Bachelder *et al.*, 2002; Belperio *et al.*, 2004; Pan *et al.*, 2006a; Phillips *et al.*, 2003), but its role in cancer angiogenesis is less clear. Previous studies have shown that stimulation of human umbilical vein endothelial cells (HUVEC) with VEGF or bFGF resulted in upregulation of cell surface expression of CXCR4 and increased migration of the cells toward CXCL12 (Salcedo *et al.*, 1999). In addition, the same investigators injected CXCL12, VEGF, or saline (control) subcutaneously in mice and evaluated the extent of microvessel formation (Salcedo *et al.*, 1999). They found that the mice that received subcutaneous injections of CXCL12 and VEGF had significant increase in microvessel formation within 4 days compared to those received saline control (Salcedo *et al.*, 1999). Using a different murine model in a study evaluating the function of CXCR4 during embryogenesis, investigators found that CXCR4 was expressed in developing vascular endothelial cells and that mice lacking CXCR4 or CXCL12 had defective formation of the large vessels supplying the gastrointestinal tract (Tachibana *et al.*, 1998). Moreover, they showed that mice lacking CXCR4 or CXCL12 died *in utero* and were defective in vascular development, hematopoesis, and cardiogenesis (Tachibana *et al.*, 1998). However, in animal models of breast, renal cell, and non-small cell lung cancer, immunoneutralization of CXCL12 or CXCR4 attenuated tumor metastases, but had no effect on the extent of angiogenesis or tumor size of the primary tumor (Muller *et al.*, 2001; Phillips *et al.*, 2003), suggesting that the CXCL12–CXCR4 biological axis mediates metastases independent of angiogenesis. These discrepant findings suggest that, although CXCL12–CXCR4 axis can mediate angiogenesis in other models, within the microenvironment of the tumor, any CXCL12 produced is most likely cleared by CXCR4 expressed by tumor cells and does not induce tumor angiogenesis.

The ELR-negative CXC chemokines, CXCL4 and CXCL14, have been shown to exhibit unique angiostatic properties. CXCL4, the original chemokine shown to inhibit angiogenesis (Maione *et al.*, 1990), signals via the receptor CXCR3. Moreover, it has been shown to inhibit angiogenesis via interaction with cell surface glycosaminoglycans or with angiogenic mediators and their receptors such as bFGF or CXCL8 (Bikfalvi and Gimenez-Gallego, 2004; Dudek *et al.*, 2003; Perollet *et al.*, 1998). CXCL4 also exists as a nonallelic gene variant, CXCL4L1, which is a very potent inhibitor of angiogenesis in both *in vitro* and *in vivo* models (Struyf *et al.*, 2004).

CXCL14 is a non-ELR CXC chemokine that was first identified in head and neck carcinoma where it was found to be downregulated in the tumor specimens (Frederick *et al.*, 2000; Hromas *et al.*, 1999; Sleeman *et al.*, 2000). Several studies support the notion that the loss or decreased expression of CXCL14 is associated with tumor formation and growth. In one study, CXCL14 was found to inhibit endothelial cell chemotaxis to CXCL8, VEGF, and bFGF *in vitro*, and to be a potent inhibitor of angiogenesis *in vivo* (Shellenberger *et al.*, 2004). In an animal model of prostate cancer, CXCL14 inhibited tumor growth when transfected into prostate cancer cells (Schwarze *et al.*, 2005). The mechanisms of action of CXCL14, as well as its receptor, remain to be elucidated.

IV. IMMUNOANGIOSTASIS

CXCR3 and its ligands contribute to antitumor defenses by two distinct mechanisms, inhibition of angiogenesis, as discussed above, and in addition, promoting Th1-dependent immunity through recruitment of CXCR3-expressing T and NK cells (Balestrieri *et al.*, 2008; Loetscher *et al.*, 1996; Luster, 1998; Qin *et al.*, 1998; Rabin *et al.*, 1999). This concept has been described as "immunoangiostasis" (Pan *et al.*, 2006b; Strieter *et al.*, 2004). In animal models of non-small cell lung cancer, investigators showed Th1 cytokine-induced cell-mediated immunity and inhibition of angiogenesis resulted in suppression of tumor growth (Hillinger *et al.*, 2003; Sharma *et al.*, 2000, 2003). In one study (Sharma *et al.*, 2000), CCL19 was shown to promote recruitment of dendritic cells and T cells and to induce a reduction in tumor size via increases in the angiostatic CXC chemokines, CXCL9 and CXCL10. In another study, the intratumor injection of CCL21 resulted in complete tumor eradication in some of the treated mice that was entirely attenuated by inhibiting CXCL9 and CXCL10, suggesting the antitumor response is secondary to CXCR3 ligands inducing a local immunoangiostatic environment (Sharma et al., 2003). In a murine model of renal cell carcinoma (Tannenbaum et al., 1998), IL-12 treatment resulted in regression of the tumor, however, this effect was lost when the CXCR3 ligands were depleted, thus, underscoring the importance of the CXCR3/CXCR3 ligand biology in tumorigenesis. The concept of immunoangiostasis has been demonstrated by investigators using a murine model of renal cell carcinoma (Pan et al., 2006b). In this study, systemic administration of IL-2 induced the expression of CXCR3 on circulating mononuclear cells, but impaired the CXCR3 chemotactic gradient from plasma to tumor by increasing circulating CXCR3 levels (Pan et al., 2006b). Systemic IL-2 administration in CXCR3$^{-/-}$ mice, however, did not inhibit tumor growth, suggesting that the antitumor effect of IL-2 was CXCR3-dependent. Additional experiments showed that the combined administration of systemic IL-2 with intratumor CXCL9 resulted in a significant reduction in tumor growth and angiogenesis, and an increase in tumor necrosis and intratumor infiltration of CXCR3+ mononuclear cells compared to treatment with IL-2 or CXCL9 alone. These results support the notion of immunoangiostasis by demonstrating the following: (1) optimization of systemic immunotherapy by combining systemic activation of mononuclear cells to express CXCR3 and enhancing the CXCR3 chemotactic gradient to promote mononuclear cell extravasation within the tumor, (2) induction of type 1 cytokine-dependent cell-mediated immunity, and (3) inhibition of tumor angiogenesis. The unique capacity to be angiostatic and to demonstrate Th1 cell-mediated immunity underscores the potential therapeutic role that the CXCR3/CXCR3 ligand axis may have in cancer.

V. CHEMOKINE-INDUCED ANGIOGENESIS IN TUMOR MODELS

Angiogenesis is essential for the growth of tumors. While much of the tumor angiogenesis research thus far has focused on the contribution of the VEGF family (Kerbel, 2008), CXC chemokine-mediated angiogenesis has been shown to play a critical role in malignancies affecting a multitude of organs including skin, pancreas, ovary, colon, stomach, lung, prostate, brain, head and neck, and kidneys. Most human malignancies express one or more of the CXC chemokine receptors including CXCR4, CXCR3, and the novel receptor for CXCL11 and CXCL12, CXCR7 (Koizumi et al., 2007).

A. Melanoma

The angiogenic chemokines, CXCL1, CXCL2, and CXCL3, were originally identified from culture supernatants of melanoma cell lines (Richmond and Thomas, 1988), and are highly expressed in human melanoma (Luan *et al.*, 1997; Owen *et al.*, 1997). Sustained transgenic expression into immortalized murine melanocytes transformed their phenotype from one that did not normally form tumors, to one that formed highly vascular tumors in immunocompetent mice. These same investigators also demonstrated that depletion of CXCL1, CXCL2, or CXCL3 in the host resulted in significant reduction in angiogenesis and tumor growth (Luan *et al.*, 1997; Owen *et al.*, 1997). The chemokine receptor, CXCR2, is also responsible for the angiogenic activity mediated by these chemokines (Addison *et al.*, 2000). A point mutation in CXCR2 results in constitutive signaling which promotes preneoplastic to neoplastic cellular transformation (Burger *et al.*, 1999). Lastly, a direct correlation between levels of CXCL8 and CXCR2, and the aggressiveness of the tumor, has been found in melanoma (Varney *et al.*, 2006).

B. Pancreatic Cancer

The ELR+ CXC chemokines and CXCR2 have been found to play important roles in human pancreatic cancer. In one study, investigators found that Capan-1, a human pancreatic cell line, expressed the angiogenic CXC chemokines CXCL1 and CXCL8 as well as CXCR2. Moreover, they found that growth of Capan-1 cells was inhibited when anti-CXCL1 or anti-CXCL8 monoclonal antibodies were added into the culture medium (Takamori *et al.*, 2000). In a separate study using a rat corneal micropocket model, investigators showed that blockade of the chemokine receptor CXCR2 inhibited pancreatic cell-induced angiogenesis (Wente *et al.*, 2006). In this study, secreted CXC chemokine levels of CXCL3, CXCL5, and CXCL8 in the supernatant of the cell lines were analyzed by ELISA; the expression of all three CXC chemokines in the supernatant of two cell lines was confirmed (Wente *et al.*, 2006). Using the corneal micropocket assay and pelleted supernatant of all three cell lines, the investigators showed that neovascularization was induced, and the angiogenesis could be significantly inhibited by the addition of anti-CXCR2 antibody (Wente *et al.*, 2006).

In a study using an orthotopic xenograft model, investigators found that constitutive NF-κB activity directly correlated with tumor angiogenesis, tumor growth, and metastases of human pancreatic cancer cells (Xiong *et al.*, 2004). In this study, blockade of the NF-κB activity significantly inhibited the *in vitro* and *in vivo* expression of VEGF and CXCL8, attenuated tumor growth, and suppressed metastases. CXCL8 has also been shown to promote the proliferation of pancreatic carcinoma cells (Kamohara *et al.*, 2007).

Recently, investigators have shown that the CXC chemokine/CXCR2 axis promotes pancreatic cancer tumor-associated angiogenesis both *in vivo* and *in vitro* (Matsuo *et al.*, 2009). Specifically, the investigators prospectively collected secretin-stimulated exocrine pancreatic secretions from normal individuals and from patients with pancreatic cancer. They found that the concentrations of the ELR+ CXC chemokines were significantly higher in patients with pancreatic cancer compared to normals (Matsuo *et al.*, 2009). Moreover, *in vitro*, they found that the ELR+ CXC chemokine levels in the supernatants from several different pancreatic cancer cell lines were significantly higher than those seen in the supernatant from a human pancreatic ductal epithelial cell line (Matsuo *et al.*, 2009). Lastly, using an orthotopic nude mouse model of pancreatic cancer, they showed significant reduction in tumor volume and microvessel density following administration of anti-CXCR2 antibody (Matsuo *et al.*, 2009).

C. Ovarian Cancer

The angiogenic CXC chemokine, CXCL8, has been implicated in the biology of ovarian carcinoma. The expression of CXCL8, VEGF, and bFGF was evaluated in human ovarian cancer cell lines (Yoneda *et al.*, 1998). In this study, all cancer cell lines expressed similar levels of bFGF *in vitro*, but levels of CXCL8 and VEGF were different across the lines (ranging from high to low expression). When implanted into the peritoneum of immunocompromised mice, the high-expressing CXCL8 tumors were associated with a significant increase in mortality (all animals died within 51 days), and the expression of CXCL8 was associated with increased tumor vascularity. In the same study, VEGF and bFGF expression were not correlated with tumor vascularity or mortality, although VEGF was associated with the presence of ascites (Yoneda *et al.*, 1998). Importantly, these findings have been validated in a separate study where investigators found that the angiogenic activity of ascites fluid from patients with ovarian cancer corresponded to levels of CXCL8 (Gawrychowski *et al.*, 1998).

D. Gastrointestinal Cancer

Several angiogenic and angiostatic CXC chemokines have been studied in cancers of the gastrointestinal tract including gastric and colorectal cancers. The angiogenic CXC chemokine, CXCL5, binds to CXCR2 and its expression has been shown to correlate with late stages of gastric cancer (Park et al., 2007). Another angiogenic ELR+ CXC chemokine, CXCL6, has been shown to be expressed by endothelial cells within gastrointestinal tumors and has been associated with neovascularization (Gijsbers et al., 2005). In one study, prostaglandin E2 was found to induce *in vivo* tumor growth by inducing expression of the angiogenic CXC chemokine, CXCL1 (Wang et al., 2006b). In another study, investigators demonstrated that high expression of CXCL12 (defined as CXCL12 positivity in 50% or more of tumor cells) in colorectal cancer cells was associated with shorter survival compared to those with low expression of CXCL12 (Akishima-Fukasawa et al., 2009). Lastly, in one study using colorectal cancer cell lines, CXCL10 significantly upregulated invasion-related properties in colorectal cancer cells by promoting matrix metalloproteinase-9 expression, adhesion to laminin, and inducing colorectal carcinoma cell migration. Their findings suggest that CXCL10 may promote progression of colorectal carcinoma (Zipin-Roitman et al., 2007).

E. Bronchogenic Cancer

The angiogenic CXC chemokines CXCL5 and CXCL8 have been shown to play important roles in lung cancer angiogenesis. In a murine model of non-small cell lung cancer (human non-small cell lung cancer/SCID mouse chimera), tumor-derived CXCL8 correlated with tumorigenesis (Arenberg, 1995), and the depletion of CXCL8 in this model was associated with a greater than 40% reduction in tumor size, and a reduction in spontaneous metastases. These findings were directly correlated to decreased tumor angiogenesis (Arenberg, 1995). Other investigators have shown that non-small cell lung cancer cell lines that constitutively express CXCL8 are more virulent, and have greater angiogenic activity (Smith *et al.*, 1994; Yatsunami *et al.*, 1997). In a study of human bronchogenic carcinoma, the angiogenic CXC chemokine, CXCL8, was shown to be directly associated with tumor angiogenesis (Smith *et al.*, 1994). These investigators found that CXCL8 levels were four times higher in human tissue homogenates of non-small cell bronchogenic carcinoma compared to normal lung tissue. Moreover, functional studies using tissue homogenates of tumors demonstrated the induction of *in vitro* endothelial chemotaxis and *in vivo* corneal neovascularization; and addition of anti-CXCL8 antibodies resulted in marked attenuation of both endothelial cell chemotaxis and neovascularization (Smith *et al.*, 1994), suggesting that tumor production of CXCL8 is critical for the neovascularization necessary for the initiation and maintenance of tumor growth.

CXCL5 has also been shown to be associated with non-small cell lung cancer angiogenesis (Arenberg *et al.*, 1998). These investigators found elevated levels of CXCL5 in human specimens of non-small cell lung cancer and that these levels were strongly correlated with the vascularity of the tumor. In a SCID mouse model of human non-small cell carcinoma, these investigators showed that expression of CXCL5 in the tumors correlated with tumor growth (Arenberg *et al.*, 1998). Passive immunization of non-small cell lung cancer tumor-bearing mice with neutralizing anti-CXCL5 antibodies reduced tumor growth, tumor vascularity and metastases, but neither the *in vitro* nor *in vivo* proliferation of non-small cell lung cancer cells was affected by CXCL5 (Arenberg *et al.*, 1998). While the lack of complete tumor inhibition is likely due to a functional redundancy among the angiogenic chemokines, overall, expression of the ELR+ CXC chemokines in human non-small cell lung cancer samples correlates with worse survival (Chen *et al.*, 2003; White *et al.*, 2003).

The common receptor for the ELR+ CXC chemokines, CXCR2, is important in mediating angiogenesis. In a murine model of heterotopic and orthotopic syngeneic Lewis lung carcinoma in C57Bl/6 mice, investigators found a correlation between the expression of endogenous ELR+ CXC chemokines and tumor growth and metastatic potential of the tumors (Keane *et al.*, 2004). In addition, tumors in the $CXCR2^{-/-}$ mice were smaller and had increased tumor necrosis, reduced vascular density, and a marked reduction in spontaneous metastases (Keane *et al.*, 2004). In a separate murine model in which $Kras^{LA1}$ mice develop spontaneous lung adenocarcinoma via somatic activation of a KRAS allele carrying an activating mutation in codon 12, elevated levels of ELR+ CXC chemokines were found in premalignant lesions of $Kras^{LA1}$ mice, and inhibition of CXCR2 blocked the expansion of early alveolar neoplastic lesions, and induced apoptosis of vascular endothelial cells within the alveolar lesions (Wislez *et al.*, 2006).

F. Prostate Cancer

In one study, investigators determined whether the expression of CXCL8 by human prostate cancer cells correlates with induction of angiogenesis, growth, and metastatic potential (Kim *et al.*, 2001). In this study, low and high CXCL8-producing cancer clones were isolated from the heterogeneous PC-3 human prostate cancer cell line. Titration studies showed that the PC-3 cells expressing high levels of CXCL8 were highly vascularized, rapidly growing, and had a 100% incidence of lymph node metastasis (Kim *et al.*, 2001). In a human/SCID mouse chimeric model of heterotopic prostate cancer, three human prostate cancer cell lines were examined for constitutive production of the ELR+ CXC chemokines (Moore *et al.*, 1999). These cancer lines were found to use different ELR+ CXC chemokines as angiogenic mediators: depletion of CXCL1, but not CXCL8, inhibited tumor growth and angiogenesis in some lines, and the converse occurred in other lines (Moore *et al.*, 1999). These findings suggest that different prostate cancers use different CXC chemokines to mediate tumor angiogenesis, a finding that has been described by other investigators (Kim *et al.*, 2001) as well as with other types of cancers (Chen *et al.*, 1999; Cohen *et al.*, 1995; Kitadai *et al.*, 1998; Miller *et al.*, 1998; Richards *et al.*, 1997; Singh *et al.*, 1994).

G. Glioblastoma

The hallmark of glioblastoma multiforme is the presence of marked angiogenesis. The mechanism underlying the angiogenesis is incompletely defined. A candidate tumor suppressor gene, ING4, was found to be down-regulated in human glioblastoma specimens (Garkavtsev *et al.*, 2004). Specimens with the lowest expression of ING4 had the greatest growth and angiogenesis when implanted into immunocompromised mice. Inhibition of CXCL8 *in vivo*, however, markedly reduced tumor-associated angiogenesis and growth of the tumor. These findings suggest a connection between ING4 and the expression of ELR+ CXC chemokines in human cancer, and may provide a unique opportunity to target ELR+ CXC chemokine-mediated angiogenesis.

H. Head and Neck Cancer

The non-ELR CXC chemokine, CXCL14, was found to be downregulated in head and neck squamous cell carcinoma specimens as compared to normal adjacent tissue (Frederick *et al.*, 2000). Subsequently, other investigators showed that CXCL14 inhibits *in vitro* endothelial cell chemotaxis in response to CXCL8, VEGF, and bFGF, and also inhibits angiogenesis in response to these mediators *in vivo* (Shellenberger *et al.*, 2004). In addition, the angiogenic CXC chemokine, CXCL5, has been associated with increased proliferation and invasion of head and neck squamous cell carcinoma thought to be, at least in part, a function of increased angiogenesis (Miyazaki *et al.*, 2006).

I. Renal Cell Cancer

In a study of patients with metastatic renal cell carcinoma, investigators evaluated tumor specimens and plasma for levels of ELR+ CXC chemokines and expression of CXCR2 (Mestas *et al.*, 2005). They found that the proangiogenic CXCR2 ligands, CXCL1, CXCL3, CXCL5, CXCL8, as well as VEGF, were elevated in the plasma and were also expressed within the tumors. Moreover, CXCR2 was expressed on endothelial cells within the tumors. These investigators also used a model of syngeneic renal cell carcinoma in BALB/c mice (Mestas *et al.*, 2005). In the CXCR2$^{-/-}$ mice, there was a marked reduction in tumor growth which correlated with decreased angiogenesis, increased tumor necrosis, and decreased metastatic potential. These findings suggest that CXCR2 and its ligands play an important role in renal cell carcinoma-associated angiogenesis and tumorigenesis.

VI. CHEMOKINES AFFECT ON CANCER METASTASES

The word "metastasis" (from the Greek for "displacement") refers to the migration of malignant cells to areas distant from the primary tumor. Tumor metastasis is an organized, organ-specific process that occurs in a stepwise fashion: (1) malignant cells are released from the primary tumor, (2) the released malignant cells invade blood vessels or lymphatics and are transported to the capillary bed of a distant organ, and (3) the malignant cells travel from the circulation to the organ parenchyma of the distant site and proliferate (Chambers *et al.*, 2002; Geiger and Peeper, 2009). Since the vast majority of cancer deaths occur due to metastasis of the tumor, rather than growth of the primary tumor, understanding this multistep process is critical in cancer biology and treatment (Kruizinga *et al.*, 2009; Leber and Efferth, 2009).

In addition to mediating cellular migration, chemokines and their receptors have been shown to affect many cellular functions including survival, adhesion, invasion, proliferation, and circulating chemokine levels. A growing body of evidence supports a chemokine-mediated mechanism for the metastatic spread of tumor cells: *in vitro* and *in vivo* models have shown that chemokines regulate tumor-associated angiogenesis (a prerequisite for metastasis), activate host tumor-specific immunologic responses, and direct tumor cell proliferation in an autocrine fashion (Ben-Baruch, 2009; Gerber *et al.*, 2009; Kruizinga *et al.*, 2009).

A. The CXCL12–CXCR4 Axis in Mediating Homing of Metastases

The ELR− CXC chemokine, CXCL12, plays an important role in stem cell motility (Hattori *et al.*, 2001) as well as tumor invasion (Chu *et al.*, 2007). While distinguishing the angiogenic activity of a chemokine from its metastatic effect may be difficult in some experimental systems, it is generally agreed that the CXCL12–CXCR4 axis plays a critical role in tumor metastases. Moreover, investigators have shown that, *in vivo*, CXCR4 is upregulated in tumor cells by the presence of hypoxia via hypoxia-inducible factor-1α (HIF-1α) (Schioppa *et al.*, 2003; Schutyser *et al.*, 2007).

The CXCL12–CXCR4 axis has been shown to be a critical factor in cancer biology in that it promotes the migration of tumor cells into metastatic sites. In fact, CXCR4 is the most common chemokine receptor that has been shown to be overexpressed in human cancer (Koizumi *et al.*, 2007). The increased expression of CXCR4 has been associated with increased metastatic potential and poor prognosis in many solid tumors, including esophageal cancer (Kaifi *et al.*, 2005; Sasaki *et al.*, 2009; Wang *et al.*, 2009), colorectal cancer (Kim *et al.*, 2005; Matsusue *et al.*, 2009; Mongan *et al.*, 2009; Speetjens *et al.*, 2009), non-small cell lung cancer (Belperio *et al.*, 2004; Oonakahara *et al.*, 2004; Phillips *et al.*, 2003; Wagner *et al.*, 2009), melanoma (Murakami *et al.*, 2004; Scala *et al.*, 2005), breast cancer (Kato *et al.*, 2003; Muller *et al.*, 2001; Smith *et al.*, 2004), ovarian cancer (Scotton *et al.*, 2002), prostate cancer (Taichman *et al.*, 2002), pancreatic cancer (Saur *et al.*, 2005), neuroblastoma (Geminder *et al.*, 2001; Russell *et al.*, 2004), osteosarcoma (Oda *et al.*, 2006), renal cell cancer (Pan et al., 2006a), and gastric cancer (Yasumoto *et al.*, 2006).

CXCL4L1, a variant of CXCL4 isolated from thrombin-stimulated platelets, has been shown to be a more potent inhibitor of endothelial cell chemotaxis compared to CXCL4 *in vitro*, and more effective than CXCL4 in inhibiting bFGF-induced angiogenesis in rat corneas (Struyf *et al.*, 2004). In a separate study using different tumor models of melanoma (B16 melanoma orthotopically propagated in C57Bl/6 mice) and lung carcinoma (A549 adenocarcinoma and Lewis lung carcinoma cell lines orthotopically propagated in C57Bl/6 and SCID mice), the same investigators showed that CXCL4L1 is a more potent inhibitor of tumor growth and metastasis than CXCL4 (Struyf *et al.*, 2007). They also demonstrated that while CXCL4L1 was more potent than CXCL10 in preventing tumor metastasis in immunocompromised mice, it had equal antitumoral activity as CXCL9 in immunocompetent mice (Struyf *et al.*, 2007). Together these data support the contention that CXCL4L1 is a highly potent antitumoral chemokine, and that it prevents the development and metastasis of tumors through its potent antiangiogenic properties.

B. CXCR7, A Novel Receptor for CXCL11 and CXCL12

While CXCL12 primarily binds to CXCR4, and CXCL11 to CXCR3, a novel receptor, CXCR7, has also been identified (Burns *et al.*, 2006). CXCR7 can regulate CXCL12-mediated migratory cues and may play a critical role in tumor cell metastases and tissue invasion (Zabel *et al.*, 2009). In a study of heterotopic transfer of human breast cancer cell line into SCID mice, and lung cancer cell line into immunocompetent mice, investigators sought to determine the role of CXCR7 in the growth of these tumors (Miao *et al.*, 2007). Using a combination of overexpression and RNA interference, they showed that CXCR7 was expressed on breast and lung cancer cell lines and promoted growth of both tumors (Miao *et al.*, 2007). In addition, they found that the expression of CXCR7 on breast cancer cells enhanced the ability of the cells to seed and proliferate in the lung, a common site of metastatic breast cancer (Miao *et al.*, 2007). Lastly, in another study using high-density tissue microarrays constructed from clinical samples from patients undergoing radical prostatectomy, CXCR7 expression was increased in tumors that were more aggressive and with increasing tumor grade (Wang *et al.*, 2008).

VII. CONCLUSION

The CXC chemokines and their receptors play a critical role in tumor angiogenesis, growth, aggressiveness, and ultimately metastasis. Understanding the mechanisms through which they work may lead to novel therapies for a wide range of human malignancies.

REFERENCES

Addison, C. L., Daniel, T. O., Burdick, M. D., Liu, H., Ehlert, J. E., Xue, Y. Y., Buechi, L., Walz, A., Richmond, A., and Strieter, R. M. (2000). The CXC chemokine receptor 2, CXCR2, is the putative receptor for ELR+ CXC chemokine-induced angiogenic activity. *J. Immunol.* **165**(9), 5269–5277.

Addison, C. L., Belperio, J. A., Burdick, M. D., and Strieter, R. M. (2004). Overexpression of the duffy antigen receptor for chemokines (DARC) by NSCLC tumor cells results in increased tumor necrosis. *BMC Cancer* **4**, 28.

Akishima-Fukasawa, Y., Nakanishi, Y., Ino, Y., Moriya, Y., Kanai, Y., and Hirohashi, S. (2009). Prognostic significance of CXCL12 expression in patients with colorectal carcinoma. *Am. J. Clin. Pathol.* **132**(2), 202–210quiz 307.

Arenberg, D. (1995). *J. Investg. Med.* **43**(Suppl. 3), 479A.

Arenberg, D. A., Keane, M. P., DiGiovine, B., Kunkel, S. L., Morris, S. B., Xue, Y. Y., Burdick, M. D., Glass, M. C., Iannettoni, M. D., and Strieter, R. M. (1998). Epithelial-neutrophil activating peptide (ENA-78) is an important angiogenic factor in non-small cell lung cancer. *J. Clin. Invest.* **102**(3), 465–472.

Bachelder, R. E., Wendt, M. A., and Mercurio, A. M. (2002). Vascular endothelial growth factor promotes breast carcinoma invasion in an autocrine manner by regulating the chemokine receptor CXCR4. *Cancer Res.* **62**(24), 7203–7206.

Balabanian, K., Lagane, B., Infantino, S., Chow, K. Y., Harriague, J., Moepps, B., Arenzana-Seisdedos, F., Thelen, M., and Bachelerie, F. (2005). The chemokine SDF-1/CXCL12 binds to and signals through the orphan receptor RDC1 in T lymphocytes. *J. Biol. Chem.* **280**(42), 35760–35766.

Balestrieri, M. L., Balestrieri, A., Mancini, F. P., and Napoli, C. (2008). Understanding the immunoangiostatic CXC chemokine network. *Cardiovasc. Res.* **78**(2), 250–256.

Belperio, J. A., Phillips, R. J., Burdick, M. D., Lutz, M., Keane, M., and Strieter, R. (2004). The SDF-1/CXCL 12/CXCR4 biological axis in non-small cell lung cancer metastases. *Chest* **125**(Suppl. 5), 156S.

Ben-Baruch, A. (2009). Site-specific metastasis formation: Chemokines as regulators of tumor cell adhesion, motility and invasion. *Cell Adh. Migr.* **3**(4), 328–333.

Bikfalvi, A., and Gimenez-Gallego, G. (2004). The control of angiogenesis and tumor invasion by platelet factor-4 and platelet factor-4-derived molecules. *Semin. Thromb. Hemost.* **30**(1), 137–144.

Burger, M., Burger, J. A., Hoch, R. C., Oades, Z., Takamori, H., and Schraufstatter, I. U. (1999). Point mutation causing constitutive signaling of CXCR2 leads to transforming activity similar to Kaposi's sarcoma herpesvirus-G protein-coupled receptor. *J. Immunol.* **163**(4), 2017–2022.

Burns, J. M., Summers, B. C., Wang, Y., Melikian, A., Berahovich, R., Miao, Z., Penfold, M. E., Sunshine, M. J., Littman, D. R., Kuo, C. J., Wei, K., McMaster, B. E., et al. (2006). A novel chemokine receptor for SDF-1 and I-TAC involved in cell survival, cell adhesion, and tumor development. *J. Exp. Med.* **203**(9), 2201–2213.

Chambers, A. F., Groom, A. C., and MacDonald, I. C. (2002). Dissemination and growth of cancer cells in metastatic sites. *Nat. Rev. Cancer* **2**(8), 563–572.

Charo, I. F., and Ransohoff, R. M. (2006). The many roles of chemokines and chemokine receptors in inflammation. *N. Engl. J. Med.* **354**(6), 610–621.

Chen, Z., Malhotra, P. S., Thomas, G. R., Ondrey, F. G., Duffey, D. C., Smith, C. W., Enamorado, I., Yeh, N. T., Kroog, G. S., Rudy, S., McCullagh, L., Mousa, S., et al. (1999). Expression of proinflammatory and proangiogenic cytokines in patients with head and neck cancer [In Process Citation]. *Clin. Cancer Res.* **5**(6), 1369–1379.

Chen, J. J., Yao, P. L., Yuan, A., Hong, T. M., Shun, C. T., Kuo, M. L., Lee, Y. C., and Yang, P. C. (2003). Up-regulation of tumor interleukin-8 expression by infiltrating macrophages: Its correlation with tumor angiogenesis and patient survival in non-small cell lung cancer. *Clin. Cancer Res.* **9**(2), 729–737.

Chu, C. Y., Cha, S. T., Chang, C. C., Hsiao, C. H., Tan, C. T., Lu, Y. C., Jee, S. H., and Kuo, M. L. (2007). Involvement of matrix metalloproteinase-13 in stromal-cell-derived factor 1 alpha-directed invasion of human basal cell carcinoma cells. *Oncogene* **26**(17), 2491–2501.

Cohen, R. F., Contrino, J., Spiro, J. D., Mann, E. A., Chen, L. L., and Kreutzer, D. L. (1995). Interleukin-8 expression by head and neck squamous cell carcinoma. *Arch. Otolaryngol. Head Neck Surg.* **121**(2), 202–209.

Du, J., Luan, J., Liu, H., Daniel, T. O., Peiper, S., Chen, T. S., Yu, Y., Horton, L. W., Nanney, L. B., Strieter, R. M., and Richmond, A. (2002). Potential role for Duffy antigen chemokine-binding protein in angiogenesis and maintenance of homeostasis in response to stress. *J. Leukoc. Biol.* **71**(1), 141–153.

Dudek, A. Z., Nesmelova, I., Mayo, K., Verfaillie, C. M., Pitchford, S., and Slungaard, A. (2003). Platelet factor 4 promotes adhesion of hematopoietic progenitor cells and binds IL-8: Novel mechanisms for modulation of hematopoiesis. *Blood* **101**(12), 4687–4694.

Ehlert, J. E., Addison, C. A., Burdick, M. D., Kunkel, S. L., and Strieter, R. M. (2004). Identification and partial characterization of a variant of human CXCR3 generated by posttranscriptional exon skipping. *J. Immunol.* **173**(10), 6234–6240.

Frederick, M. J., Henderson, Y., Xu, X., Deavers, M. T., Sahin, A. A., Wu, H., Lewis, D. E., El-Naggar, A. K., and Clayman, G. L. (2000). In vivo expression of the novel CXC chemokine BRAK in normal and cancerous human tissue. *Am. J. Pathol.* **156**(6), 1937–1950.

Garkavtsev, I., Kozin, S. V., Chernova, O., Xu, L., Winkler, F., Brown, E., Barnett, G. H., and Jain, R. K. (2004). The candidate tumour suppressor protein ING4 regulates brain tumour growth and angiogenesis. *Nature* **428**(6980), 328–332.

Gawrychowski, K., Skopinska-Rozewska, E., Barcz, E., Sommer, E., Szaniawska, B., Roszkowska-Purska, K., Janik, P., and Zielinski, J. (1998). Angiogenic activity and interleukin-8 content of human ovarian cancer ascites. *Eur. J. Gynaecol. Oncol.* **19**(3), 262–264.

Geiger, T. R., and Peeper, D. S. (2009). Metastasis mechanisms. *Biochim. Biophys. Acta* **1796**(2), 293–308.

Geminder, H., Sagi-Assif, O., Goldberg, L., Meshel, T., Rechavi, G., Witz, I. P., and Ben-Baruch, A. (2001). A possible role for CXCR4 and its ligand, the CXC chemokine stromal cell-derived factor-1, in the development of bone marrow metastases in neuroblastoma. *J. Immunol.* **167**(8), 4747–4757.

Gerber, P. A., Hippe, A., Buhren, B. A., Muller, A., and Homey, B. (2009). Chemokines in tumor-associated angiogenesis. *Biol. Chem.* **390**, 1213–1223.

Gijsbers, K., Gouwy, M., Struyf, S., Wuyts, A., Proost, P., Opdenakker, G., Penninckx, F., Ectors, N., Geboes, K., and Van Damme, J. (2005). GCP-2/CXCL6 synergizes with other endothelial cell-derived chemokines in neutrophil mobilization and is associated with angiogenesis in gastrointestinal tumors. *Exp. Cell Res.* **303**(2), 331–342.

Hattori, K., Heissig, B., Tashiro, K., Honjo, T., Tateno, M., Shieh, J. H., Hackett, N. R., Quitoriano, M. S., Crystal, R. G., Rafii, S., and Moore, M. A. (2001). Plasma elevation of stromal cell-derived factor-1 induces mobilization of mature and immature hematopoietic progenitor and stem cells. *Blood* **97**(11), 3354–3360.

Heidemann, J., Ogawa, H., Dwinell, M. B., Rafiee, P., Maaser, C., Gockel, H. R., Otterson, M. F., Ota, D. M., Lugering, N., Domschke, W., and Binion, D. G. (2003). Angiogenic effects of interleukin 8 (CXCL8) in human intestinal microvascular endothelial cells are mediated by CXCR2. *J. Biol. Chem.* **278**(10), 8508–8515.

Hillinger, S., Yang, S. C., Zhu, L., Huang, M., Duckett, R., Atianzar, K., Batra, R. K., Strieter, R. M., Dubinett, S. M., and Sharma, S. (2003). EBV-induced molecule 1 ligand chemokine (ELC/CCL19) promotes IFN-gamma-dependent antitumor responses in a lung cancer model. *J. Immunol.* **171**(12), 6457–6465.

Hromas, R., Broxmeyer, H. E., Kim, C., Nakshatri, H., Christopherson, K., 2nd, Azam, M., and Hou, Y. H. (1999). Cloning of BRAK, a novel divergent CXC chemokine preferentially expressed in normal versus malignant cells. *Biochem. Biophys. Res. Commun.* **255**(3), 703–706.

Kaifi, J. T., Yekebas, E. F., Schurr, P., Obonyo, T., Wachowiak, R., Busch, P., Heinecke, A., Pantel, K., and Izbicki, J. R. (2005). Tumor-cell homing to lymph nodes and bone marrow and CXCR4 expression in esophageal cancer. *J. Natl. Cancer Inst.* **97**(24), 1840–1847.

Kamohara, H., Takahashi, M., Ishiko, T., Ogawa, M., and Baba, H. (2007). Induction of interleukin-8 (CXCL-8) by tumor necrosis factor-alpha and leukemia inhibitory factor in pancreatic carcinoma cells: Impact of CXCL-8 as an autocrine growth factor. *Int. J. Oncol.* **31**(3), 627–632.

Kato, M., Kitayama, J., Kazama, S., and Nagawa, H. (2003). Expression pattern of CXC chemokine receptor-4 is correlated with lymph node metastasis in human invasive ductal carcinoma. *Breast Cancer Res.* **5**(5), R144–150.

Keane, M. P., Belperio, J. A., Xue, Y. Y., Burdick, M. D., and Strieter, R. M. (2004). Depletion of CXCR2 inhibits tumor growth and angiogenesis in a murine model of lung cancer. *J. Immunol.* **172**(5), 2853–2860.

Keeley, E. C., Mehrad, B., and Strieter, R. M. (2008). Chemokines as mediators of neovascularization. *Arterioscler. Thromb. Vasc. Biol.* **28**(11), 1928–1936.

Kerbel, R. S. (2008). Tumor angiogenesis. *N. Engl. J. Med.* **358**(19), 2039–2049.

Kim, S. J., Uehara, H., Karashima, T., McCarty, M., Shih, N., and Fidler, I. J. (2001). Expression of interleukin-8 correlates with angiogenesis, tumorigenicity, and metastasis of human prostate cancer cells implanted orthotopically in nude mice. *Neoplasia* **3**(1), 33–42.

Kim, J., Takeuchi, H., Lam, S. T., Turner, R. R., Wang, H. J., Kuo, C., Foshag, L., Bilchik, A. J., and Hoon, D. S. (2005). Chemokine receptor CXCR4 expression in colorectal cancer patients increases the risk for recurrence and for poor survival. *J. Clin. Oncol.* **23**(12), 2744–2753.

Kitadai, Y., Haruma, K., Sumii, K., Yamamoto, S., Ue, T., Yokozaki, H., Yasui, W., Ohmoto, Y., Kajiyama, G., Fidler, I. J., and Tahara, E. (1998). Expression of interleukin-8 correlates with vascularity in human gastric carcinomas. *Am. J. Pathol.* **152**(1), 93–100.

Koizumi, K., Hojo, S., Akashi, T., Yasumoto, K., and Saiki, I. (2007). Chemokine receptors in cancer metastasis and cancer cell-derived chemokines in host immune response. *Cancer Sci.* **98**(11), 1652–1658.

Kruizinga, R. C., Bestebroer, J., Berghuis, P., de Haas, C. J., Links, T. P., de Vries, E. G., and Walenkamp, A. M. (2009). Role of chemokines and their receptors in cancer. *Curr. Pharm. Des.* **15**(29), 3396–3416.

Lasagni, L., Francalanci, M., Annunziato, F., Lazzeri, E., Giannini, S., Cosmi, L., Sagrinati, C., Mazzinghi, B., Orlando, C., Maggi, E., Marra, F., Romagnani, S., *et al.* (2003). An alternatively spliced variant of CXCR3 mediates the inhibition of endothelial cell growth induced by IP-10, Mig, and I-TAC, and acts as functional receptor for platelet factor 4. *J. Exp. Med.* **197**(11), 1537–1549.

Leber, M. F., and Efferth, T. (2009). Molecular principles of cancer invasion and metastasis (review). *Int. J. Oncol.* **34**(4), 881–895.

Loetscher, M., Gerber, B., Loetscher, P., Jones, S. A., Piali, L., Clark-Lewis, I., Baggiolini, M., and Moser, B. (1996). Chemokine receptor specific for IP10 and mig: Structure, function, and expression in activated T-lymphocytes. *J. Exp. Med.* **184**(3), 963–969.

Loetscher, M., Loetscher, P., Brass, N., Meese, E., and Moser, B. (1998). Lymphocyte-specific chemokine receptor CXCR3: Regulation, chemokine binding and gene localization. *Eur. J. Immunol.* **28**(11), 3696–3705.

Luan, J., Shattuck-Brandt, R., Haghnegahdar, H., Owen, J. D., Strieter, R., Burdick, M., Nirodi, C., Beauchamp, D., Johnson, K. N., and Richmond, A. (1997). Mechanism and biological significance of constitutive expression of MGSA/GRO chemokines in malignant melanoma tumor progression. *J. Leukoc. Biol.* **62**(5), 588–597.

Luster, A. D. (1998). Chemokines—Chemotactic cytokines that mediate inflammation. *N. Engl. J. Med.* **338**(7), 436–445.

Luster, A. D., Cardiff, R. D., MacLean, J. A., Crowe, K., and Granstein, R. D. (1998). Delayed wound healing and disorganized neovascularization in transgenic mice expressing the IP-10 chemokine. *Proc. Assoc. Am. Physicians* **110**(3), 183–196.

Maione, T. E., Gray, G. S., Petro, J., Hunt, A. J., Donner, A. L., Bauer, S. I., Carson, H. F., and Sharpe, R. J. (1990). Inhibition of angiogenesis by recombinant human platelet factor-4 and related peptides. *Science* **247**(4938), 77–79.

Matsuo, Y., Raimondo, M., Woodward, T. A., Wallace, M. B., Gill, K. R., Tong, Z., Burdick, M. D., Yang, Z., Strieter, R. M., Hoffman, R. M., and Guha, S. (2009). CXC-chemokine/CXCR2 biological axis promotes angiogenesis in vitro and in vivo in pancreatic cancer. *Int. J. Cancer* **125**(5), 1027–1037.

Matsusue, R., Kubo, H., Hisamori, S., Okoshi, K., Takagi, H., Hida, K., Nakano, K., Itami, A., Kawada, K., Nagayama, S., and Sakai, Y. (2009). Hepatic stellate cells promote liver metastasis of colon cancer cells by the action of SDF-1/CXCR4 axis. *Ann. Surg. Oncol.* **16**(9), 2645–2653.

Mehrad, B., Keane, M. P., and Strieter, R. M. (2007). Chemokines as mediators of angiogenesis. *Thromb. Haemost.* **97**(5), 755–762.

Mestas, J., Burdick, M. D., Reckamp, K., Pantuck, A., Figlin, R. A., and Strieter, R. M. (2005). The role of CXCR2/CXCR2 ligand biological axis in renal cell carcinoma. *J. Immunol.* **175**(8), 5351–5357.

Miao, Z., Luker, K. E., Summers, B. C., Berahovich, R., Bhojani, M. S., Rehemtulla, A., Kleer, C. G., Essner, J. J., Nasevicius, A., Luker, G. D., Howard, M. C., and Schall, T. J. (2007). CXCR7 (RDC1) promotes breast and lung tumor growth in vivo and is expressed on tumor-associated vasculature. *Proc. Natl. Acad. Sci. USA* **104**(40), 15735–15740.

Miller, L. J., Kurtzman, S. H., Wang, Y., Anderson, K. H., Lindquist, R. R., and Kreutzer, D. L. (1998). Expression of interleukin-8 receptors on tumor cells and vascular endothelial cells in human breast cancer tissue. *Anticancer Res.* **18**(1A), 77–81.

Miyazaki, H., Patel, V., Wang, H., Edmunds, R. K., Gutkind, J. S., and Yeudall, W. A. (2006). Down-regulation of CXCL5 inhibits squamous carcinogenesis. *Cancer Res.* **66**(8), 4279–4284.

Mongan, J. P., Fadul, C. E., Cole, B. F., Zaki, B. I., Suriawinata, A. A., Ripple, G. H., Tosteson, T. D., and Pipas, J. M. (2009). Brain metastases from colorectal cancer: Risk factors, incidence, and the possible role of chemokines. *Clin. Colorectal Cancer* **8**(2), 100–105.

Moore, B. B., Arenberg, D. A., Stoy, K., Morgan, T., Addison, C. L., Morris, S. B., Glass, M., Wilke, C., Xue, Y. Y., Sitterding, S., Kunkel, S. L., Burdick, M. D., *et al.* (1999). Distinct CXC chemokines mediate tumorigenicity of prostate cancer cells. *Am. J. Pathol.* **154**(5), 1503–1512.

Muller, A., Homey, B., Soto, H., Ge, N., Catron, D., Buchanan, M. E., McClanahan, T., Murphy, E., Yuan, W., Wagner, S. N., Barrera, J. L., Mohar, A., *et al.* (2001). Involvement of chemokine receptors in breast cancer metastasis. *Nature* **410**(6824), 50–56.

Murakami, T., Cardones, A. R., and Hwang, S. T. (2004). Chemokine receptors and melanoma metastasis. *J. Dermatol. Sci.* **36**(2), 71–78.

Murdoch, C., Monk, P. N., and Finn, A. (1999). Cxc chemokine receptor expression on human endothelial cells. *Cytokine* **11**(9), 704–712.

Oda, Y., Yamamoto, H., Tamiya, S., Matsuda, S., Tanaka, K., Yokoyama, R., Iwamoto, Y., and Tsuneyoshi, M. (2006). CXCR4 and VEGF expression in the primary site and the metastatic site of human osteosarcoma: Analysis within a group of patients, all of whom developed lung metastasis. *Mod. Pathol.* **19**(5), 738–745.

Oonakahara, K., Matsuyama, W., Higashimoto, I., Kawabata, M., Arimura, K., and Osame, M. (2004). Stromal-derived factor-1alpha/CXCL12-CXCR 4 axis is involved in the dissemination of NSCLC cells into pleural space. *Am. J. Respir. Cell Mol. Biol.* **30**(5), 671–677.

Owen, J. D., Strieter, R., Burdick, M., Haghnegahdar, H., Nanney, L., Shattuck-Brandt, R., and Richmond, A. (1997). Enhanced tumor-forming capacity for immortalized melanocytes expressing melanoma growth stimulatory activity/growth-regulated cytokine beta and gamma proteins. *Int. J. Cancer* **73**(1), 94–103.

Pan, J., Mestas, J., Burdick, M. D., Phillips, R. J., Thomas, G. V., Reckamp, K., Belperio, J. A., and Strieter, R. M. (2006a). Stromal derived factor-1 (SDF-1/CXCL12) and CXCR4 in renal cell carcinoma metastasis. *Mol. Cancer* **5**, 56.

Pan, J., Burdick, M. D., Belperio, J. A., Xue, Y. Y., Gerard, C., Sharma, S., Dubinett, S. M., and Strieter, R. M. (2006b). CXCR3/CXCR3 ligand biological axis impairs RENCA tumor growth by a mechanism of immunoangiostasis. *J. Immunol.* **176**(3), 1456–1464.

Park, J. Y., Park, K. H., Bang, S., Kim, M. H., Lee, J. E., Gang, J., Koh, S. S., and Song, S. Y. (2007). CXCL5 overexpression is associated with late stage gastric cancer. *J. Cancer Res. Clin. Oncol.* **133**(11), 835–840.

Perollet, C., Han, Z. C., Savona, C., Caen, J. P., and Bikfalvi, A. (1998). Platelet factor 4 modulates fibroblast growth factor 2 (FGF-2) activity and inhibits FGF-2 dimerization. *Blood* **91**(9), 3289–3299.

Phillips, R. J., Burdick, M. D., Lutz, M., Belperio, J. A., Keane, M. P., and Strieter, R. M. (2003). The stromal derived factor-1/CXCL12-CXC chemokine receptor 4 biological axis in non-small cell lung cancer metastases. *Am. J. Respir. Crit. Care Med.* **167**(12), 1676–1686.

Qin, S., Rottman, J. B., Myers, P., Kassam, N., Weinblatt, M., Loetscher, M., Koch, A. E., Moser, B., and Mackay, C. R. (1998). The chemokine receptors CXCR3 and CCR5 mark subsets of T cells associated with certain inflammatory reactions. *J. Clin. Invest.* **101**(4), 746–754.

Rabin, R. L., Park, M. K., Liao, F., Swofford, R., Stephany, D., and Farber, J. M. (1999). Chemokine receptor responses on T cells are achieved through regulation of both receptor expression and signaling. *J. Immunol.* **162**(7), 3840–3850.

Richards, B. L., Eisma, R. J., Spiro, J. D., Lindquist, R. L., and Kreutzer, D. L. (1997). Coexpression of interleukin-8 receptors in head and neck squamous cell carcinoma. *Am. J. Surg.* **174**(5), 507–512.

Richmond, A., and Thomas, H. G. (1988). Melanoma growth stimulatory activity: Isolation from human melanoma tumors and characterization of tissue distribution. *J. Cell. Biochem.* **36**(2), 185–198.

Rollins, B. J. (1997). Chemokines. *Blood* **90**(3), 909–928.

Russell, H. V., Hicks, J., Okcu, M. F., and Nuchtern, J. G. (2004). CXCR4 expression in neuroblastoma primary tumors is associated with clinical presentation of bone and bone marrow metastases. *J. Pediatr. Surg.* **39**(10), 1506–1511.

Salcedo, R., Wasserman, K., Young, H. A., Grimm, M. C., Howard, O. M., Anver, M. R., Kleinman, H. K., Murphy, W. J., and Oppenheim, J. J. (1999). Vascular endothelial growth factor and basic fibroblast growth factor induce expression of CXCR4 on human endothelial cells: In vivo neovascularization induced by stromal-derived factor-1alpha. *Am. J. Pathol.* **154**(4), 1125–1135.

Sasaki, K., Natsugoe, S., Ishigami, S., Matsumoto, M., Okumura, H., Setoyama, T., Uchikado, Y., Kita, Y., Tamotsu, K., Hanazono, K., Owaki, T., and Aikou, T. (2009). Expression of CXCL12 and its receptor CXCR4 in esophageal squamous cell carcinoma. *Oncol. Rep.* **21**(1), 65–71.

Saur, D., Seidler, B., Schneider, G., Algul, H., Beck, R., Senekowitsch-Schmidtke, R., Schwaiger, M., and Schmid, R. M. (2005). CXCR4 expression increases liver and lung metastasis in a mouse model of pancreatic cancer. *Gastroenterology* **129**(4), 1237–1250.

Scala, S., Ottaiano, A., Ascierto, P. A., Cavalli, M., Simeone, E., Giuliano, P., Napolitano, M., Franco, R., Botti, G., and Castello, G. (2005). Expression of CXCR4 predicts poor prognosis in patients with malignant melanoma. *Clin. Cancer Res.* **11**(5), 1835–1841.

Schioppa, T., Uranchimeg, B., Saccani, A., Biswas, S. K., Doni, A., Rapisarda, A., Bernasconi, S., Saccani, S., Nebuloni, M., Vago, L., Mantovani, A., Melillo, G., et al. (2003). Regulation of the chemokine receptor CXCR4 by hypoxia. *J. Exp. Med.* **198**(9), 1391–1402.

Schutyser, E., Su, Y., Yu, Y., Gouwy, M., Zaja-Milatovic, S., Van Damme, J., and Richmond, A. (2007). Hypoxia enhances CXCR4 expression in human microvascular endothelial cells and human melanoma cells. *Eur. Cytokine Netw.* **18**(2), 59–70.

Schwarze, S. R., Luo, J., Isaacs, W. B., and Jarrard, D. F. (2005). Modulation of CXCL14 (BRAK) expression in prostate cancer. *Prostate* **64**(1), 67–74.

Scotton, C. J., Wilson, J. L., Scott, K., Stamp, G., Wilbanks, G. D., Fricker, S., Bridger, G., and Balkwill, F. R. (2002). Multiple actions of the chemokine CXCL12 on epithelial tumor cells in human ovarian cancer. *Cancer Res.* **62**(20), 5930–5938.

Sharma, S., Stolina, M., Luo, J., Strieter, R. M., Burdick, M., Zhu, L. X., Batra, R. K., and Dubinett, S. M. (2000). Secondary lymphoid tissue chemokine mediates T cell-dependent antitumor responses in vivo. *J. Immunol.* **164**(9), 4558–4563.

Sharma, S., Yang, S. C., Hillinger, S., Zhu, L. X., Huang, M., Batra, R. K., Lin, J. F., Burdick, M. D., Strieter, R. M., and Dubinett, S. M. (2003). SLC/CCL21-mediated antitumor responses require IFNgamma, MIG/CXCL9 and IP-10/CXCL10. *Mol. Cancer* **2**, 22.

Shellenberger, T. D., Wang, M., Gujrati, M., Jayakumar, A., Strieter, R. M., Burdick, M. D., Ioannides, C. G., Efferson, C. L., El-Naggar, A. K., Roberts, D., Clayman, G. L., and Frederick, M. J. (2004). BRAK/CXCL14 is a potent inhibitor of angiogenesis and a chemotactic factor for immature dendritic cells. *Cancer Res.* **64**(22), 8262–8270.

Shen, H., Schuster, R., Stringer, K. F., Waltz, S. E., and Lentsch, A. B. (2006). The Duffy antigen/receptor for chemokines (DARC) regulates prostate tumor growth. *FASEB J.* **20**(1), 59–64.

Singh, R. K., Gutman, M., Radinsky, R., Bucana, C. D., and Fidler, I. J. (1994). Expression of interleukin 8 correlates with the metastatic potential of human melanoma cells in nude mice. *Cancer Res.* **54**(12), 3242–3247.

Sleeman, M. A., Fraser, J. K., Murison, J. G., Kelly, S. L., Prestidge, R. L., Palmer, D. J., Watson, J. D., and Kumble, K. D. (2000). B cell- and monocyte-activating chemokine (BMAC), a novel non-ELR alpha-chemokine. *Int. Immunol.* **12**(5), 677–689.

Smith, D. R., Polverini, P. J., Kunkel, S. L., Orringer, M. B., Whyte, R. I., Burdick, M. D., Wilke, C. A., and Strieter, R. M. (1994). Inhibition of interleukin 8 attenuates angiogenesis in bronchogenic carcinoma. *J. Exp. Med.* **179**(5), 1409–1415.

Smith, M. C., Luker, K. E., Garbow, J. R., Prior, J. L., Jackson, E., Piwnica-Worms, D., and Luker, G. D. (2004). CXCR4 regulates growth of both primary and metastatic breast cancer. *Cancer Res.* **64**(23), 8604–8612.

Speetjens, F. M., Liefers, G. J., Korbee, C. J., Mesker, W. E., van de Velde, C. J., van Vlierberghe, R. L., Morreau, H., Tollenaar, R. A., and Kuppen, P. J. (2009). Nuclear localization of CXCR4 determines prognosis for colorectal cancer patients. *Cancer Microenviron.* **2**(1), 1–7.

Strieter, R. M., Polverini, P. J., Kunkel, S. L., Arenberg, D. A., Burdick, M. D., Kasper, J., Dzuiba, J., Van Damme, J., Walz, A., Marriott, D., *et al.* (1995). The functional role of the ELR motif in CXC chemokine-mediated angiogenesis. *J. Biol. Chem.* **270**(45), 27348–27357.

Strieter, R. M., Belperio, J. A., Burdick, M. D., Sharma, S., Dubinett, S. M., and Keane, M. P. (2004). CXC chemokines: Angiogenesis, immunoangiostasis, and metastases in lung cancer. *Ann. N. Y. Acad. Sci.* **1028**, 351–360.

Struyf, S., Burdick, M. D., Proost, P., Van Damme, J., and Strieter, R. M. (2004). Platelets release CXCL4L1, a nonallelic variant of the chemokine platelet factor-4/CXCL4 and potent inhibitor of angiogenesis. *Circ. Res.* **95**(9), 855–857.

Struyf, S., Burdick, M. D., Peeters, E., Van den Broeck, K., Dillen, C., Proost, P., Van Damme, J., and Strieter, R. M. (2007). Platelet factor-4 variant chemokine CXCL4L1 inhibits melanoma and lung carcinoma growth and metastasis by preventing angiogenesis. *Cancer Res.* **67**(12), 5940–5948.

Tachibana, K., Hirota, S., Iizasa, H., Yoshida, H., Kawabata, K., Kataoka, Y., Kitamura, Y., Matsushima, K., Yoshida, N., Nishikawa, S., Kishimoto, T., and Nagasawa, T. (1998). The chemokine receptor CXCR4 is essential for vascularization of the gastrointestinal tract. *Nature* **393**(6685), 591–594.

Taichman, R. S., Cooper, C., Keller, E. T., Pienta, K. J., Taichman, N. S., and McCauley, L. K. (2002). Use of the stromal cell-derived factor-1/CXCR4 pathway in prostate cancer metastasis to bone. *Cancer Res.* **62**(6), 1832–1837.

Takamori, H., Oades, Z. G., Hoch, O. C., Burger, M., and Schraufstatter, I. U. (2000). Autocrine growth effect of IL-8 and GROalpha on a human pancreatic cancer cell line, Capan-1. *Pancreas* **21**(1), 52–56.

Tannenbaum, C. S., Tubbs, R., Armstrong, D., Finke, J. H., Bukowski, R. M., and Hamilton, T. A. (1998). The CXC chemokines IP-10 and Mig are necessary for IL-12-mediated regression of the mouse RENCA tumor. *J. Immunol.* **161**(2), 927–932.

Vandercappellen, J., Van Damme, J., and Struyf, S. (2008). The role of CXC chemokines and their receptors in cancer. *Cancer Lett.* **267**(2), 226–244.

Varney, M. L., Johansson, S. L., and Singh, R. K. (2006). Distinct expression of CXCL8 and its receptors CXCR1 and CXCR2 and their association with vessel density and aggressiveness in malignant melanoma. *Am. J. Clin. Pathol.* **125**(2), 209–216.

Wagner, P. L., Hyjek, E., Vazquez, M. F., Meherally, D., Liu, Y. F., Chadwick, P. A., Rengifo, T., Sica, G. L., Port, J. L., Lee, P. C., Paul, S., Altorki, N. K., *et al.* (2009). CXCL12 and CXCR4 in adenocarcinoma of the lung: Association with metastasis and survival. *J. Thorac. Cardiovasc. Surg.* **137**(3), 615–621.

Wang, J., Ou, Z. L., Hou, Y. F., Luo, J. M., Shen, Z. Z., Ding, J., and Shao, Z. M. (2006a). Enhanced expression of Duffy antigen receptor for chemokines by breast cancer cells attenuates growth and metastasis potential. *Oncogene* **25**(54), 7201–7211.

Wang, D., Wang, H., Brown, J., Daikoku, T., Ning, W., Shi, Q., Richmond, A., Strieter, R., Dey, S. K., and DuBois, R. N. (2006b). CXCL1 induced by prostaglandin E2 promotes angiogenesis in colorectal cancer. *J. Exp. Med.* **203**(4), 941–951.

Wang, J., Shiozawa, Y., Wang, J., Jung, Y., Pienta, K. J., Mehra, R., Loberg, R., and Taichman, R. S. (2008). The role of CXCR7/RDC1 as a chemokine receptor for CXCL12/SDF-1 in prostate cancer. *J. Biol. Chem.* **283**(7), 4283–4294.

Wang, D. F., Lou, N., Zeng, C. G., Zhang, X., and Chen, F. J. (2009). Expression of CXCL12/CXCR4 and its correlation to prognosis in esophageal squamous cell carcinoma. *Chin. J. Cancer* **28**(2), 154–158.

Wente, M. N., Keane, M. P., Burdick, M. D., Friess, H., Buchler, M. W., Ceyhan, G. O., Reber, H. A., Strieter, R. M., and Hines, O. J. (2006). Blockade of the chemokine receptor CXCR2 inhibits pancreatic cancer cell-induced angiogenesis. *Cancer Lett.* **241**(2), 221–227.

White, E. S., Flaherty, K. R., Carskadon, S., Brant, A., Iannettoni, M. D., Yee, J., Orringer, M. B., and Arenberg, D. A. (2003). Macrophage migration inhibitory factor and CXC chemokine expression in non-small cell lung cancer: Role in angiogenesis and prognosis. *Clin. Cancer Res.* **9**(2), 853–860.

Wislez, M., Fujimoto, N., Izzo, J. G., Hanna, A. E., Cody, D. D., Langley, R. R., Tang, H., Burdick, M. D., Sato, M., Minna, J. D., Mao, L., Wistuba, I., *et al.* (2006). High expression of ligands for chemokine receptor CXCR2 in alveolar epithelial neoplasia induced by oncogenic kras. *Cancer Res.* **66**(8), 4198–4207.

Xiong, H. Q., Abbruzzese, J. L., Lin, E., Wang, L., Zheng, L., and Xie, K. (2004). NF-kappaB activity blockade impairs the angiogenic potential of human pancreatic cancer cells. *Int. J. Cancer* **108**(2), 181–188.

Yasumoto, K., Koizumi, K., Kawashima, A., Saitoh, Y., Arita, Y., Shinohara, K., Minami, T., Nakayama, T., Sakurai, H., Takahashi, Y., Yoshie, O., and Saiki, I. (2006). Role of the CXCL12/CXCR4 axis in peritoneal carcinomatosis of gastric cancer. *Cancer Res.* **66**(4), 2181–2187.

Yatsunami, J., Tsuruta, N., Ogata, K., Wakamatsu, K., Takayama, K., Kawasaki, M., Nakanishi, Y., Hara, N., and Hayashi, S. (1997). Interleukin-8 participates in angiogenesis in non-small cell, but not small cell carcinoma of the lung. *Cancer Lett.* **120**(1), 101–108.

Yoneda, J., Kuniyasu, H., Crispens, M. A., Price, J. E., Bucana, C. D., and Fidler, I. J. (1998). Expression of angiogenesis-related genes and progression of human ovarian carcinomas in nude mice. *J. Natl. Cancer Inst.* **90**(6), 447–454.

Zabel, B. A., Wang, Y., Lewen, S., Berahovich, R. D., Penfold, M. E., Zhang, P., Powers, J., Summers, B. C., Miao, Z., Zhao, B., Jalili, A., Janowska-Wieczorek, A., *et al.* (2009). Elucidation of CXCR7-mediated signaling events and inhibition of CXCR4-mediated tumor cell transendothelial migration by CXCR7 ligands. *J. Immunol.* **183**(5), 3204–3211.

Zipin-Roitman, A., Meshel, T., Sagi-Assif, O., Shalmon, B., Avivi, C., Pfeffer, R. M., Witz, I. P., and Ben-Baruch, A. (2007). CXCL10 promotes invasion-related properties in human colorectal carcinoma cells. *Cancer Res.* **67**(7), 3396–3405.

Genetically Engineered Mouse Models in Cancer Research

Jessica C. Walrath, Jessica J. Hawes, Terry Van Dyke, and Karlyne M. Reilly

Mouse Cancer Genetics Program, National Cancer Institute
Frederick, Maryland, USA

I. Introduction
II. Generation of Genetically Engineered Mouse Models
   A. Studying Loss of Gene Function in Mouse Models
   B. Studying Gain of Gene Function in Mouse Models
   C. Modeling Chromosomal Translocations in Mice
   D. Studying the Mechanisms and Timing of Tumorigenesis
   E. Discovery of Novel Tumor Genes
III. Cancer Paradigms and Lessons from the Mouse
   A. The Cancer Stem Cell and Initiating Cell
   B. The Tumor Microenvironment
IV. Summary
   References

Mouse models of human cancer have played a vital role in understanding tumorigenesis and answering experimental questions that other systems cannot address. Advances continue to be made that allow better understanding of the mechanisms of tumor development, and therefore the identification of better therapeutic and diagnostic strategies. We review major advances that have been made in modeling cancer in the mouse and specific areas of research that have been explored with mouse models. For example, although there are differences between mice and humans, new models are able to more accurately model sporadic human cancers by specifically controlling timing and location of mutations, even within single cells. As hypotheses are developed in human and cell culture systems, engineered mice provide the most tractable and accurate test of their validity *in vivo*. For example, largely through the use of these models, the microenvironment has been established to play a critical role in tumorigenesis, since tumor development and the interaction with surrounding stroma can be studied as both evolve. These mouse models have specifically fueled our understanding of cancer initiation, immune system roles, tumor angiogenesis, invasion, and metastasis, and the relevance of molecular diversity observed among human cancers. Currently, these models are being designed to facilitate *in vivo* imaging to track both primary and metastatic tumor development from much earlier stages than previously possible. Finally, the approaches developed in this field to achieve basic understanding are emerging as effective tools to guide much needed development of treatment strategies, diagnostic strategies, and patient stratification strategies in clinical research.

I. INTRODUCTION

In 2008, for the first time in 10 years of reporting, the incidence and death rates from cancer declined significantly (Jemal et al., 2008). Although encouraging, there were still just under 1.5 million new cases of cancer in the United States, and just over 500,000 deaths per year. The decrease is attributed to a reduction in the most common cancers, lung, prostate, colorectal, and breast, likely due to better screening for early detection of cancer and reduced tobacco use (Jemal et al., 2008). Among the remaining cancers, rates remained stable or increased, highlighting the need for continued research to develop better prevention, detection, and treatment of cancer.

Although improvements in prevention and early detection are leading to a drop in cancer incidence and death, the ability to treat established cancers is still quite limited. Relatively new therapies, such as Gleevec, bevacizumab, and herceptin, are beginning to make a difference to patients with particular molecular subtypes of cancer, demonstrating the power of correctly targeted molecular-based therapy (Normanno et al., 2009; Soverini et al., 2008; Tan et al., 2008), but also highlighting the limitation in our understanding of the molecular pathways critical to cancer and how they vary between individuals. Cancer is a complex disease in which normal cellular pathways are altered to give rise to the properties leading to cancer, such as inappropriate growth, survival, and invasion. While the study of human tumors has yielded many insights into the molecular changes present in cancers, more rigorous testing of hypotheses through experimental manipulation is necessary to better understand which changes are causative, and therefore targetable, and which are secondary. Mouse model systems provide an experimentally tractable mammalian system to test the hypotheses generated from the observation and study of human tumors, and also provide opportunities to identify novel mechanisms to be confirmed in human tumors. Cross-species comparison has proven to be powerful in improving the understanding of a wide variety of human diseases, including cancer (e.g., see Brown et al., 2008; Kim et al., 2005; Wang and Paigen, 2005; Wang et al., 2005). We describe in this chapter the engineering tools and methods currently available to model cancer in the mouse, and highlight how some key challenges have been and are being addressed using these sophisticated approaches.

Humans are a highly heterogeneous population with variation in genetic background, diet, and environmental exposures. Any of these variables can affect how cancer manifests itself, and thus can become confounding issues in the analysis of human tumors. Studies in mice allow researchers to control these variables to simplify experiments and ask well-structured questions. For example, genetic background can be held constant while testing the effects of diet or carcinogens on cancer, and diet/environment can be held constant while examining the effects of genetic changes on tumorigenesis. As mammals, mice share many anatomical, cellular, and molecular traits with humans that are known to have critical functions in cancer, such as an immune system, maternal effects in utero, imprinting of genes, and alternative splicing. Thus, mice provide an experimentally tractable model system with a wealth of developed research tools to understand basic mechanisms of cancer.

One of the difficulties and frustrations in the development of cancer therapies is that preclinical studies have historically had limited predictive value for the efficacy of drugs in patients. Preclinical studies have been relatively predictive of toxicity in humans, such that many drugs that enter Phase I clinical trials are taken forward to Phase II clinical trials due to tolerable toxicity profiles. In contrast, most drugs (95%) are dropped in Phase II and Phase III clinical trials because of limited efficacy against cancer (e.g., see Sharpless and Depinho, 2006). There are several potential explanations for this. It is clear that some drugs are only effective in a subset of patients (Haas-Kogan et al., 2005; Mellinghoff et al., 2005), but because current preclinical testing strategies do not model the relevant molecular and cellular milieu, drug specificity for subtypes can go unappreciated. Indeed, preclinical vetting of drugs is largely performed on xenografts of human tumor cell lines grown subcutaneously in immune-compromised mice. It is now clear, however, that cancer stroma is distinct from normal stroma and provides a supportive microenvironment for tumor growth. Human cancer xenografts lack the appropriate cancer stroma support when grown over relatively short periods of time in immune-compromised mice. As a result, they may be more vulnerable to drug treatment than cancers that have coevolved with their stroma (Olive et al., 2009). Similarly, inflammatory responses play an integral role in cancer development (de Visser and Coussens, 2006; Tlsty and Coussens, 2006), but are not effectively modeled in xenografts. Moreover, immune-compromised mice could mount a limited immune attack on engrafted cells (Quintana et al., 2008), making them more susceptible to treatment.

Mice engineered with cell-specific human cancer-associated aberrations can address many of these issues, as tumors arise *de novo* in the context of a normal immune system and coevolve with surrounding stroma. Furthermore, more than 100 years of genetic research in the mouse has provided powerful strategies for assessing the complex genetics of disease susceptibility, therapeutic responses, and associated toxicities that are so diverse among the human population. Identifying these parameters is key to the development of personalized medicine; while current technologies have accelerated the pace of human epidemiological association studies, there is no doubt that discovery can be accelerated by cross-species comparisons. Finally, several initiatives are under way to use the *de novo* mouse cancer models in preclinical guidance for human clinical trials. While challenging due to the complexity of the human diseases and the difficulty of modeling them accurately in mice, early studies indicate that this approach offers great promise, opening up a new era of translational research using engineered mouse models (Becher and Holland, 2006; Carver and Pandolfi, 2006; Frese and Tuveson, 2007; Gutmann *et al.*, 2006; Huse and Holland, 2009; Pritchard *et al.*, 2003; Sharpless and Depinho, 2006).

While subcutaneous xenograft models have not been predictive for targeted cancer therapies in humans, they continue to provide a simple testing ground for triage and pharmacodynamic studies prior to analysis in more complex models. However, of significance, xenograft approaches refined to better model the tumor microenvironment may hold significant promise in predicting some human therapeutic outcomes and are currently under study. These methods include the transplantation of human primary tumors into immunocompromised mice at the orthotopic site and into mice that have been "humanized" with transplanted human hematopoietic and other organ systems (reviewed in Bibby, 2004; Legrand *et al.*, 2009; Richmond and Su, 2008; Sharpless and Depinho, 2006). In the end, it is likely that a series of integrated model systems will hold the most power in discovery and prediction of effective human disease management. The uses of xenograft models have been reviewed elsewhere (Lee *et al.*, 2007; Pegram and Ngo, 2006; Sausville and Burger, 2006; Teicher, 2009; Troiani *et al.*, 2008), and we focus in this chapter on genetically engineered mouse models of cancer. We describe here the state of the art techniques for modeling cancer in mice both for basic mechanistic research and for preclinical applied studies. We also review the issues in cancer research for which these mouse models are particularly useful due to the preservation of temporal disease development, including coevolution of cancer and stromal compartments.

II. GENERATION OF GENETICALLY ENGINEERED MOUSE MODELS

With the availability of the complete sequence of the mouse genome, technology to manipulate the mouse genome, and well-defined inbred strains as well as extensive information on the polymorphisms among strains, the ability to engineer mice to test hypotheses of tumorigenesis is impressive. Experiments can now be easily undertaken to assess the outcome when the function of a gene is lost, mutated, underexpressed, or overexpressed in the appropriate cell types *in vivo*. In addition to individual research efforts, several organized initiatives have been launched using a variety of approaches to systematically target every gene in the genome (Adams *et al*., 2004; Friedel *et al*., 2007; Hansen *et al*., 2008; Schnutgen *et al*., 2005; Skarnes *et al*., 2004; To *et al*., 2004). Additionally, much can be learned about the natural history of cancer through the use of reporters to image molecular, cellular, and anatomical changes during tumorigenesis. Finally, mutagenesis studies in the mouse have identified new cancer-causing mutations that can then be confirmed in studies of human cancers.

A. Studying Loss of Gene Function in Mouse Models

1. MOUSE GENE KNOCKOUTS

Studying the loss of function of genes provides insight into understanding the biological functions for which the protein product is required. Loss-of-function studies most commonly use "knockout" strategies to remove the gene of interest by engineering constitutive or conditional deletions in the gene. For genes that span large genomic regions, deletion of the first few exons encoding the start codon is often sufficient to block transcription or translation into a functional protein product. However, sometimes an alternative start codon or alternative splicing can lead to a truncated protein product with partial function that can mask the significance of the gene of interest in biological loss-of-function studies. Thus, careful molecular characterization of genetically engineered alleles is important to verify that the function of the gene is truly lost and also that additional inadvertent gene rearrangements or deletions are not present.

In translational cancer research, loss-of-function studies provide a powerful approach to assessing the potential validity of targeted therapies, since the target can be specifically inactivated in the context of a developing or developed tumor. In this approach, accurate interpretation requires understanding of any functional redundancy differences between mouse and human. For example, if one species can utilize alternative gene products in a cancer pathway, but the other cannot, results will not be concordant. Fortunately, careful study using existing technologies, along with the available extensive genomic information, allows for this assessment and provides additional information on which to base further therapeutic design.

In basic research studies, the use of knockout strategies have been critical in understanding cause and effect relationships in cancer development, and can be applied to the assessment of many gene classes, including oncogenes, tumor suppressor genes, and metabolic ("housekeeping") genes. The classification of a gene as a tumor suppressor depends on the demonstration that impaired function can facilitate tumor development, and that can only be achieved in the context of a developing tumor within the host. Germline loss-of-function (homozygous deletion) studies often lead to embryonic lethality precluding assessment for adult diseases, making conclusive determination of tumor suppressor genes more difficult. In many cancer models, however, animals heterozygous for a tumor suppressor knockout allele are susceptible to tumor formation either due to haploinsufficiency (impaired function from insufficient levels) or by somatic loss of the wild type allele (Cichowski *et al.*, 1999; Kost-Alimova and Imreh, 2007; Macleod and Jacks, 1999; Reilly *et al.*, 2000; Zheng and Lee, 2002). Nonetheless, a more valid and versatile approach is to conditionally induce a somatic mutation, as discussed below. Use of loss-of-function mouse models to study cancer has been extensively reviewed elsewhere (Maddison and Clarke, 2005; Van Dyke and Jacks, 2002). Knockout gene targeting strategies and techniques have also been reviewed extensively (Capecchi, 2005; LePage and Conlon, 2006; Mikkola and Orkin, 2005; Porret *et al.*, 2006) and will not be discussed in detail here.

2. MOUSE CONDITIONAL GENE MUTATIONS

With conventional knockouts, loss of a vital gene can often lead to embryonic lethality, severe developmental abnormalities, or adult sterility, making it impossible to study the gene in the desired disease context. For example, one cannot study the role of genes such as BRCA1 and BRCA2 in breast cancer if the animals die before birth or adulthood (Evers and Jonkers, 2006). In addition, ablation of the gene of interest in the entire body does not mimic spontaneous tumorigenesis in humans, where tumors evolve in a wild-type environment, and the timing of gene loss may be a critical factor in disease development.

To circumvent conventional knockout limitations, sophisticated conditional genetic engineering technology has been developed to create systems where genetic events can be tightly controlled spatially and temporally. Bacterial Cre and yeast FLP enzymes are site-specific recombinases that catalyze specific recombination between defined 34 bp *loxP* and *FRT* sites, respectively (Branda and Dymecki, 2004). Therefore, in the presence of Cre or FLP protein expression, homologous recombination is induced between *loxP* or *FRT* sites that flank the gene of interest and are oriented in the same direction, thus recombining out the flanked genetic sequence and deleting the gene of interest. By temporally and spatially controlling expression of the recombinase, it is then possible to temporally and spatially control deletion of the gene of interest, overcoming interference from developmental abnormalities and lethality (Branda and Dymecki, 2004). Mice carrying the Cre or FLP recombinase under control of a tissue-specific promoter are crossed with mice carrying the gene of interest flanked by *loxP* or *FRT* sites to conditionally knockout the gene in a specific tissue or cell type, such as progenitor cells, or at specific times during development.

Multiple types of Cre delivery systems have been developed to temporally and spatially control Cre expression. These include promoter-driven cell or tissue-specific (discussed above), viral, somatically introduced, and temporally inducible systems. Viral Cre, such as adenoviral or lentiviral Cre, in which the *Cre* gene is packaged into viral particles, can be locally delivered topically or by injection to infect cells and create a regional or clonal knockout of cells within a given area (Jackson *et al.*, 2001; Marumoto *et al.*, 2009). Both adeno- and lentiviral vectors have advantages and disadvantages that should be considered in the context of the experiment (DuPage *et al.*, 2009). For example, adenoviral vectors typically induce a significant inflammatory response that can confound interpretations of causation. However, adenoviral infection is transient and does not result in viral genome insertion. Lentiviruses are suitable for infecting nondividing somatic cells while adenovirus requires dividing cells. Lentiviral genome integration can also present a confounding variable due to insertion site mutation, although there is evidence that this may be less of an issue with lentivirus than with other insertional viruses (Gonzalez-Murillo *et al.*, 2008; Montini *et al.*, 2009). Nonetheless, the ability to mark a clone by a specific insertion site or long-term expression of a viral gene product also offers a strategy to track tumorigenesis (De Palma *et al.*, 2003; Hosoda *et al.*, 2009).

Conditional Cre systems that temporally control Cre expression (e.g., using the tet promoter or ER fusion systems described below) are continuing to be developed and offer the greatest amount of control over gene removal. Two of the most commonly used systems are the tetracycline-inducible system (Gossen and Bujard, 1992) and the tamoxifen-inducible system (Metzger and Chambon, 2001) (see also Bockamp *et al.*, 2008, and references therein). The tetracycline-repressor-based system is composed of a transactivator and an effector. The DNA-binding domain of the *Escherichia coli tetR* gene fused to the transactivation domain of the herpes simplex virion protein 16 (VP16) gene (tetR/VP16) makes up the tetracycline-controlled transactivator (tTA) that can then be driven by tissue-specific promoters (Baron *et al.*, 1997). The tTA binds to the tetracycline operator (tetO) that controls the activity of the human cytomegalovirus promoter driving conditional gene expression, including Cre to generate conditional knockouts. In this Tet-Off system, tTA is bound by tetracycline, or its more stable analogue doxycycline, inhibiting association with the tetO and blocking gene transcription. In the Tet-On system, the tetracycline-repressor has been mutated (rtTA) such that it is only in the proper conformation for association with tetO when it is bound to tetracycline or doxycycline, thus inducing expression of the gene in the presence of drug. The expression of tTA or rtTA can lead to cytotoxicity, likely due to overexpression of the

VP16 transactivator domain and titration of transcription factor machinery (Berger et al., 1990), which has spurred the development of next-generation tTAs (Bockamp et al., 2008). Success using the tetracycline-inducible systems depends on screening transgenic mouse lines for tight repression, good induction, and appropriate levels of repressor expression to avoid secondary toxicity issues.

The tamoxifen-inducible system depends on fusion of the Cre recombinase gene to a mutated ligand-binding domain of the human estrogen receptor (Cre-ER(T)) that is specifically activated by tamoxifen. In the absence of tamoxifen, the ER fusion protein is excluded from the nucleus, but is transported to the nucleus upon binding to tamoxifen where Cre can then recombine DNA. Using these techniques, temporal expression of Cre can be manually controlled by simply delivering or withholding tamoxifen. Other, less commonly used inducible systems include the insect steroid molting hormone ecdysone (No et al., 1996), the progesterone antagonist mifepristone (Ngan et al., 2002), the Lac operator–repressor (Cronin et al., 2001), and the GAL4/UAS system (Wang et al., 1999). While all of these inducible systems have been shown to be functional in the mouse, they differ in their efficiency and leakiness that affects tight control of gene expression. Because recombinase-mediated excision of the gene is irreversible, leaky Cre expression can have substantial and permanent effects that could compromise experimental outcomes.

3. MOUSE MODELS OF RNA INTERFERENCE

Alternatively, loss-of-function studies can use RNA interference (RNAi) to specifically knockdown the expression of target genes posttranscriptionally before the mRNA can be translated into protein (Meister and Tuschl, 2004). Target sequence-specific small interfering RNAs (siRNAs) are short antisense peptides 21–28 nucleotides long. The RNA-induced silencing complex (RISC) recognizes the double-stranded siRNA fragments and cleaves the endogenous complementary messenger RNA (mRNA) that is then rapidly degraded. Though siRNA has been widely used to knockdown expression of target genes, it is limited to transient transfection *in vitro*. Short hairpin RNA (shRNA) can be used to cause long-term gene knockdown, both *in vitro* and *in vivo*. The shRNAs are much longer, generally between 50 and 70 nucleotides in length, allowing them to be transcribed as stable messages *in vivo*. They are made as a single-strand molecule that then forms a short hairpin tertiary structure, folding in on itself to form a stem–loop structure *in vivo*. After transcription and folding, the enzyme Dicer cleaves off the loop leaving behind a double-stranded siRNA molecule that can then be recognized by RISC.

The shRNA sequences can be cloned into viral vectors to be stably incorporated into the genome for sustained knockdown of target genes.

Several groups have used shRNA-expressing constructs to successfully create RNAi transgenic knockdown animals (Sandy et al., 2005). The shRNA expressing mice can be generated by inserting the promoter-shRNA construct into the mouse genome using normal knockout and transgenic techniques, including embryonic stem cell electroporation as well as lentiviral shRNA transgene injection into the pronucleus of fertilized eggs or the perivitelline space of single-cell mouse embryos. Knockin shRNA animals can also be made using homologous recombination to target the insertion of the shRNA transgene to a specific site, such as the ubiquitously expressed *Rosa26* locus. Since shRNAs are promoter-driven, they can be constructed using inducible, reversible, or tissue-specific promoters (Dickins et al., 2007), or with lox-STOP-lox (LSL)(described in Section II.B.3), or other Cre-lox-regulated control elements (Tiscornia et al., 2004; Ventura et al., 2004). Recently, Cre-regulated RNAi was achieved by combining a Cre-regulated shRNA transgene within an FLP expression cassette such that the RNAi transgene and a green fluorescent protein (GFP) reporter are inverted upon recombination to induce expression (Stern et al., 2008). Although these methods have shown promise, they do have limitations, especially regarding the incomplete penetrance of the silencing, possibly due to varying expression of shRNAs from retroviral constructs (reviewed in Westbrook et al., 2005).

As the details of RNAi in mammalian cells became better understood, new shRNAs have been developed that take advantage of natural miRNA biogenesis (Silva et al., 2005; Westbrook et al., 2005). Artificial vectors are designed that contain the shRNA sequence within miR30 (Silva et al., 2005), a naturally occurring, well-characterized microRNA transcript (Cullen, 2004; Zeng et al., 2002). The expression of mature shRNA and subsequent knockdown of the target gene are greater when expressed from the endogenous miR30 promoter than from exogenous promoters, due to more efficient microRNA biogenesis (Westbrook et al., 2005). Current studies are focusing on utilizing lentiviral vectors to deliver shRNA-miRs and generate *in vivo* mRNA knockdown (Singer and Verma, 2008; Stegmeier et al., 2005).

Since RNAi results in a knockdown of gene expression rather than complete inhibition, RNAi transgenic animals are likely to be hypomorphs as opposed to null phenocopies. RNAi is also susceptible to nonspecific effects and off-target repression of other genes. Nevertheless, RNAi animal models are a powerful approach to studying gene knockdown, because targeting constructs can be rapidly designed and tested, genes that give a lethal phenotype when knocked out can be studied without needing the complex genetic crosses involved in conditional knockouts, and model organisms can be studied for which knockout technology has not been developed, such as the worm and rat.

4. MOUSE SINGLE-CELL KNOCKOUTS

Human cancers are spontaneous diseases caused by accumulation of somatic mutations that arise from single mutant cells. Sporadic tumors developing in a wild-type or heterozygous microenvironment can develop differently compared to those developing in a homogeneous genetically mutant environment. Therefore, conventional knockout, knockin, and transgenic GEM tumor models, even tissue-specific or cell-specific models, in which the deletion or mutation occurs in a whole organ or animal, are limited in that they cannot reproduce the clonal nature of human cancer (Fig. 1). Tumors from heterozygous mouse models, such as $p53^{+/-}$ mice, arise from clonal expansion of cells that have sporadically lost the wild-type allele of the tumor suppressor. While these loss of heterozygosity-dependent models more closely recapitulate the clonal nature of human tumors, the incidence and spectrum of spontaneous tumor development is dependent on variables that affect the stochastic nature of the loss of heterozygosity event (Attardi and Donehower, 2005). Single-cell knockout approaches can be used to control tumor-initiating events by driving genetic events in individual cells, generating homozygous mutant cells on a heterozygous mutant background at very low frequency to create *in vivo* mosaics (reviewed in Lozano and Behringer, 2007). The clonal nature of single-cell knockouts can be used to more clearly elucidate the mechanisms of cellular processes such as cell cycle control and tumor formation in the context of a more normal microenvironment, potentially redefining the roles of genes previously studied in other GEM models.

Sporadic single-cell knockout mouse models have been created using genetic techniques derived from *Drosophila* systems. Recombinase enzymes

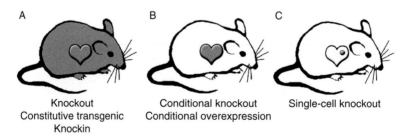

Fig. 1 Genetically engineered mouse models of cancer. Technology is now available to control the overexpression and loss of gene expression, shown in gray, (A) throughout the mouse, as is often the case with classic knockouts, constitutive transgenics, or in some cases knockins; (B) in a specific organ (e.g., the heart) or at a specific time in development, using conditional knockouts or conditional overexpression transgenics or knockins; and (C) in single-cell knockouts through sporadic loss events.

can be used under tissue specific and temporal controls to induce rare G2 mitotic interchromosomal recombination events between *FRT* or *loxP* sites residing on homologous mouse chromosomes, yielding a wild-type and a mutant daughter cell amidst a population of heterozygous cells. During the G2 phase of the cell cycle, homologous chromosomes align in preparation for mitosis. While aligned, the Cre or FLP recombinase can recombine the homologues such that regions of the chromosome are homozygosed in the daughter cells. For example, Wang *et al.* (2007) used a ubiquitously expressed recombinase to induce rare interchromosomal recombination events between *FRT* or *loxP* sites centromeric to *p53* and generate single cell $p53^{-/-}$ clones (Wang *et al.*, 2007) within a $p53^{+/-}$ environment. The sporadic *p53* knockout phenotype more closely resembles human Li–Fraumeni syndrome, in which the human *p53* gene is mutated, than previous models. This system can be used to target loss of heterozygosity events to particular cells or tissues using different Cre or FLP recombinase mouse lines.

The mosaic analysis with double markers (MADM) approach described by Muzumdar *et al.*(2007) uses a tissue-specific Cre/loxP system to induce sporadic mitotic interchromosomal recombination and knockout target genes while simultaneously labeling resultant daughter cells (Muzumdar *et al.*, 2007). Homozygous knockout and wild-type daughter cells are identified with red and green markers and further distinguished from neighboring yellow heterozygous cells. Thus, MADM allows progeny from single loss of heterozygosity events to be tracked and distinguished from normal cells within the microenvironment.

B. Studying Gain of Gene Function in Mouse Models

1. MOUSE CONSTITUTIVE TRANSGENIC MODELS

Gain-of-function studies are often used to study oncogenes in mouse models. Transgenic or knockin animals constitutively overexpressing an oncogene can be used to study how the oncogene drives tumorigenesis *in vivo*. Transgenic animals have been very useful in studying many oncogenes and for creating Cre-expressing mice and other inducible systems. Transgenic animals are created by the pronuclear injection of transgenes directly into fertilized oocytes, followed by implantation into pseudopregnant females (Macleod and Jacks, 1999; Porret *et al.*, 2006). The transgene is randomly incorporated into the genome and thus can incorporate into a gene necessary for development or fertility, causing deleterious effects and limiting the usefulness of the transgenic model. Furthermore, the epigenetic regulation of gene expression in the region surrounding the transgene integration can affect transgene expression levels and often result in silencing. Therefore, multiple founders must be screened to confirm adequate and specific expression of the transgene.

In addition to insertion site effects, transgenes often do not show the expression patterns of the endogenous gene promoter, due to the lack of regulatory sequences that can be located several kilobases from the coding region of the gene of interest. Bacterial artificial chromosomes (BAC) contain large fragments of genomic DNA (150–300 kb), and therefore can accommodate entire genes, including the cis-regulatory sequences, allowing for increased fidelity of gene expression (Shizuya et al., 1992). Under control of the gene's regulatory elements, the gene of interest will likely show the endogenous expression pattern (Swing and Sharan, 2004). Due to their large size, BACs are more likely to insert at a lower copy number, increasing the chances of single copy transgenics, which will be expressed at endogenous levels due to the regulatory sequences. Although the effects of additional genes on the BAC must be considered, molecular analysis of the BAC can be used to eliminate the possibility of other known genes being present, as well as confirm the presence of the entire coding sequence of the gene of interest (Swing and Sharan, 2004). With the development of recombineering technology (Sharan et al., 2009), large BACs can be rapidly modified and used to generate transgenic mice to compare the phenotypic effects of precise changes in a gene (Chang et al., 2009) or regulatory sequence.

2. MOUSE GENE KNOCKIN MODELS

To circumvent transgenic limitations associated with random insertion, knockin mice are created by inserting a gene of interest into a specific region of the genome using homologous recombination techniques, much like those used when creating knockouts. The *Rosa26* locus is commonly used as an insertion site for knockin animals because it is devoid of essential genes and allows for good expression of the transgene (Friedrich and Soriano, 1991). While transgenics have the potential for multiple insertions of the transgene, knockin animals carry only one copy of the transgene. Knockin approaches can also be used to replace a normal gene copy with a mutated version of the gene to examine specific mutational events in the context of normal control of the gene (e.g., Johnson et al., 2001; Lang et al., 2004).

3. MOUSE CONDITIONAL OVEREXPRESSION MODELS

Transgenic and knockin expression of deleterious genes may lead to lethality, sterility, and developmental defects that impede study of the gene of interest in cancer, as seen with many conventional knockouts. Therefore, spatial and temporal control of transgene and knockin expression may be necessary to circumvent these limitations. Conditional transgenics and conditional knockins can be created using tissue-specific promoters to constitutively drive expression or created by inserting a strong translational and transcriptional termination (STOP) sequence flanked by *loxP* or *FRT* sites in between the promoter sequence and the gene of interest (Lakso *et al.*, 1992). Examples of commonly used STOP cassettes are the lox-STOP-lox (LSL) in which multiple STOP sequences are arrayed between *loxP* sites (Jackson *et al.*, 2001) and the NEO-STOP cassette in which the neomycin resistance gene and a STOP sequence is inserted between *loxP* sites (Dragatsis and Zeitlin, 2001). The presence of the STOP sequence blocks transcription of the gene of interest. However, in the presence of Cre or FLP recombinase, the STOP cassette is removed, allowing expression. Since gene expression is dependent on excision of the STOP cassette and recombinase expression, gene expression can be spatially, temporally, and inducibly controlled with the myriad of Cre systems. STOP cassettes can be leaky and thus must be assessed empirically. STOP technology has greatly expanded the opportunities for controlling gene expression in gain-of-function studies. For example, initial K-*rasG12D* transgenic models of lung carcinoma showed a large variance in tumor penetrance, as well as death due to respiratory failure prior to tumor progression (Johnson *et al.*, 2001). Treatment of the conditional *LSL-K-ras G12D* strain with virally-infected allowed control of the timing, location, and number of tumors, therefore allowing the study of initiation and early-stage pulmonary adenocarcinoma (Jackson *et al.*, 2001).

C. Modeling Chromosomal Translocations in Mice

1. TRANSLOCATOR MICE

Human cancers, such as leukemia, often involve chromosomal translocations that lead to fusion proteins with new functions that play a role in tumorigenesis. Mouse models have been developed to study *de novo* protein functions as a result of chromosomal translocations. Basic approaches such as transgenic or knockin insertion of fusion genes resulting from human chromosomal translocations can lead to embryonic lethality or nonauthentic phenotypes (Prosser and Bradley, 2003). While this can be circumvented with conditional transgenic or conditional knockin approaches, these approaches still suffer from the limitation that they produce a population of cells expressing the fusion gene within the specified tissue compartment and do not recapitulate the clonal onset of the disease in humans. Translocation-derived cancers develop from a single cell that undergoes chromosomal translocation and then clonally expands in an environment that is initially composed of normal cells. Therefore, a more physiological approach to mimic chromosomal translocations in human disease is through induction of the chromosomal translocation event in mice (Forster *et al.*, 2003, 2005a; Prosser and Bradley, 2003). This can be accomplished by inserting *loxP* sites in trans at chromosomal breakpoints observed in human cancers, resulting in Cre-dependent interchromosomal reciprocal translocation (Forster *et al.*, 2003; van der Weyden *et al.*, 2009) (Fig. 2). This is similar to the approach used for single-cell knockouts, described above, except that the *loxP* sites are on different chromosomes, rather than homologous chromosomes. As with single-cell knockouts, the interchromosomal recombination is a rare event and as such more closely mimics the situation seen in human cancers. If the chromosomal translocation results in a fusion protein with *de novo* function animals may develop cancers similar to those in humans. For example in a mouse model of the MLL-ENL translocation fusion protein, the animals can develop a leukemia very similar to humans, with high penetrance and most likely from a clonal origin (Forster *et al.*, 2003). Since not all mouse models using this approach successfully develop cancer (Buchholz *et al.*, 2000; Collins *et al.*, 2000), it is very likely that cancer development resulting from Cre-driven interchromosomal reciprocal translocations are dependent on the expression level and distribution of the Cre transgene expression (Forster *et al.*, 2003). This could reflect the importance of the translocations occurring in the proper cell type, compartment and/or fusion gene sequence for cancer development. The success of this approach also depends on the compatibility of the transcriptional orientation of the two mouse chromosome genes.

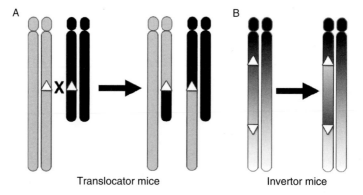

Fig. 2 Modeling chromosomal rearrangements in mouse models of cancer. (A) Interchromosomal rearrangements are generated in translocator mice by inserting single flox recombination sites on different chromosomes. Recombination between the sites exchanges pieces of the two chromosomes. (B) Intrachromosomal rearrangements are generated by inserting two flox recombination sites on the same chromosome oriented in opposite directions. By using flox sites with two different mutations, the resulting flox sites on the inverted chromosome cannot recombine and the inversion becomes stable in the presence of Cre enzyme.

2. INVERTOR MICE

Since not all fusion genes arising from chromosomal translocations in human cancers are compatible with the Cre-driven interchromosomal translocation approach, a conditional invertor approach has been used to study the *de novo* function of fusion genes in mice (Forster *et al.*, 2005b; Lobato *et al.*, 2008). Chromosomal inversions occur when two breaks occur within the same chromosome and the genetic material in between the breaks is inverted. This is achieved using *loxP* sites orientated in opposite directions at the breakpoint sites resulting in inversion of the flanked DNA sequence containing the second half of the fusion protein (Fig. 2). Thus, after inversion, the sequence for the entire fusion protein is in frame allowing for expression of the *de novo* protein (Forster *et al.*, 2005b).

Since recombination products between inverted and wild-type chromosomes do not produce viable gametes, inverted regions suppress recombination. Therefore, inversions can also be used as genetic tools to maintain the linkage of mutant loci on chromosomes. Inverted regions carrying phenotypic markers, such as the dominant *K14 Agouti* coat-color gene, are known as balancer chromosomes and can be used for large scale mutagenesis screens, similar to experiments in *Drosophila* (van der Weyden *et al.*, 2009).

D. Studying the Mechanisms and Timing of Tumorigenesis

1. MOLECULAR-GENETIC IMAGING

Molecular-genetic imaging utilizes reporter transgenes to visualize proteomic, metabolic, cellular, or genetic events *in vivo*. The three major methods of animal imaging are optical, magnetic resonance imaging, and nuclear medicine (for review see Kang and Chung, 2008; Lyons, 2005). Although conventional magnetic resonance imaging, computed tomography, ultrasound, and radiography imaging modalities have the advantage of not being dependent on genetically engineered reporter mice, they are limited to visualization of anatomic morphological changes. Optical imaging using genetically engineered reporters can be used to detect rapid molecular and genetic changes, including those that occur in response to drug treatment or during early stages of carcinogenesis before the onset of anatomical change. Furthermore, fluorescent and bioluminescent optical imaging modalities are sensitive, less expensive, less time consuming, and more convenient and user friendly than other imaging techniques.

Reporter transgenes can be made to express reporter genes under the control of specific promoter/enhancer elements to assess promoter activation *in vivo* during tumorigenesis or drug treatment. For example, transgenic reporter mice expressing luciferase under control of the estrogen receptor enhancer have been used to study the dynamics of estrogen receptor activity in response to various ligands and drug treatments (Maggi *et al.*, 2004). Furthermore, transgenic reporter mice can be crossed onto mouse models of human cancer to visualize tumor growth *in vivo*. For example, transgenic mice expressing firefly luciferase under control of the human E2F1 promoter, which is active during proliferation, can be crossed with mouse cancer models to characterize tumor cell proliferation *in vivo* (Momota and Holland, 2005).

Firefly luciferase is the most commonly used reporter gene for bioluminescence in mice, although renilla luciferase has some advantages in that it does not require ATP, and so may be less dependent on changes in the metabolic state of cells. Bioluminescence can be detected within 10 min of intraperitoneal injection of the luciferase substrate luciferin and multiple animals can be imaged together leading to a relatively quick turn-around time. Although bioluminescence has the advantage of extremely low background signal and high sensitivity, photon emissions are attenuated by hair color and the amount of tissue between the tumor and the detector, making it difficult to accurately compare tumors at varying depths and in different coat color backgrounds.

Green fluorescent protein (GFP) and red fluorescent protein (RFP) are commonly used as reporters for *in vivo* imaging (Hoffman, 2009). Since there are five different fluorescent proteins with unique emission wavelengths between 450 and 650 nm, simultaneous imaging of multiple fluorescent reporters in one animal is possible. Since fluorescence requires excitation, tissue scattering and absorption by endogenous fluorochromes can limit the sensitivity of *in vivo* fluorescence imaging.

2. RECOMBINATION REPORTERS AND LINEAGE TRACING

Cre- and FLP-dependent reporters are useful not only for determining where site-specific recombination of conditional knockout and transgenes occur (Soriano, 1999) but also for lineage tracing studies (Chai *et al.*, 2000; Jiang *et al.*, 2000). A ubiquitous promoter drives a conditional reporter transgene, such as a *lacZ* or GFP reporter, and is interrupted by the presence of a STOP cassette flanked by *loxP* or *FRT* sites. When conditional reporter transgenic mice are crossed with site-specific Cre or FLP mice driven by cell-type specific promoters, the STOP cassette of the conditional transgene is removed only in Cre- or FLP-expressing cells and the reporter continues to be expressed in the progeny of the recombined cells. For example, by using mice carrying a transgene for Cre or FLP expression in progenitor cells, the reporter expression is turned on in the progenitor cells and inherited by all subsequent progeny.

Yamamoto *et al.* (2009) have recently described a sophisticated dual reporter $R26^{NZG}$ mouse line for Cre- and FLP-dependent lineage analysis. The $R26^{NZG}$ reporter mice were constructed by knocking the NZG reporter cassette into the *Rosa26* locus. The NZG reporter cassette consists of the CAG promoter followed by the neo-STOP cassette, the *lacZ* reporter gene flanked by *FRT* sites, and finally the enhanced GFP (EGFP) gene. In the absence of Cre and FLP recombinases, neither reporter is expressed. In the presence of Cre recombinase, the *PGK-neo* cassette is removed and *lacZ* expression is turned on; however, EGFP expression is still turned off due to a STOP codon in the preceding *lacZ* sequence. In the subsequent presence of FLP recombinase, the *lacZ* sequence is removed and EGFP is expressed. Thus, *lacZ* expression is turned on following Cre-expression and EGFP expression is only turned on after expression of both Cre and FLP. Although there is an increasing breadth of Cre-expressing lines becoming available, the advantages of the dual reporter technology are currently limited by the smaller availability and lower efficiency of FLP-expressing lines. Nevertheless, the $R26^{NZG}$ mouse line in conjunction with mouse cancer models is likely to be a powerful tool in future lineage-tracing studies of tumor initiating cells.

3. GENOMIC INSTABILITY

Human cancers often show abnormal karyotypes that include, most prevalently, aneuploidy and multiple chromosomal translocations. Genomic instability is believed to contribute to human cancer by increasing the number and rate of genetic changes seen in a tumor, thus accelerating cancer development. However, the identification of the "driver" mutations necessary for tumorigenesis is confounded by the presence of many "passenger" alterations that do not contribute to tumorigenesis. Because mouse models of cancer are predisposed to tumorigenesis by engineered mutations, they can result in tumors with far fewer chromosomal aberrations, giving a means to separate driver and passenger mutations through cross-species comparisons (Peeper and Berns, 2006). For example, by comparing copy number aberrations found in mouse and human melanomas and liver tumors, new oncogenes associated with chromosomal aberrations were identified (Kim *et al.*, 2006; Zender *et al.*, 2006). Between mouse models, differences in genome stability can correlate to differences in tumor subtype, as shown in mammary tumor models with either *p53* loss or *p53* and *Brca1* loss (Liu *et al.*, 2007). Both tumor models develop mammary tumors; however, the *p53;Brca1* null tumors show greatly increased genomic instability and more closely resemble human *BRCA1* mutant tumors molecularly, whereas the *p53* null tumors show very few chromosomal alterations and more closely resemble human estrogen receptor mutant breast cancer.

While there are advantages to the reduced mutation number seen in many mouse tumors, the mutations driving mouse tumors serve to limit the range of mutations that develop. In this way, specific mouse models may filter out human "driver" mutations that are important in a different genetic context, in addition to filtering out the "passenger" mutations in cross-species comparisons. To ensure a more unbiased approach, mice with increased genomic instability may be better models for comparative oncogenomics (Maser *et al.*, 2007). Multiple approaches have been developed to drive chromosomal instability in mouse models and better mimic the structural abnormalities seen in human tumors.

In humans, loss of telomeres as a part of aging is believed to contribute to cancer incidence due to genomic changes and chromosomal fusions. Mouse chromosomes have much longer telomeres than humans and sustained telomerase expression. This may explain the reduced number of chromosomal abnormalities in mouse tumors, as well as the difference in tumor types seen, as epithelial cancers are believed to require more aberrations than the mesenchymal tumors that are more commonly seen in mice. In order to examine the importance of telomeres in cancer development, telomerase knockout mice were generated (Lee *et al.*, 1998). Early generations of telomerase-deficient mice show initial resistance to tumorigenesis, due to apoptosis or cell-cycle arrest (Chin *et al.*, 1999; Karlseder *et al.*, 1999). As mice are bred over successive generations, the long mouse telomeres erode in the absence of telomerase, and in late generations tumorigenesis occurs through increased chromosome instability. Interestingly, mice with both telomere deficiency and loss of p53 show both an increased rate of tumorigenesis and a shift in tumor spectrum toward the epithelial cancers seen in humans (Artandi *et al.*, 2000). Additionally, the chromosomal changes seen in these mouse epithelial cancers mimic those often seen in human tumors, suggesting that increasing genomic instability contributes toward a more accurate model of human tumor development (O'Hagan *et al.*, 2002).

Another system to investigate genomic instability is the mouse model of Bloom syndrome, a genetic disorder involving mutation of the *BLM* helicase gene. Bloom syndrome is associated with both genomic instability and increased risk of a wide variety of malignancies (German, 1993). A conditional mouse model of *Blm* results in mice with an increased rate of mitotic recombination, leading to an increased loss of heterozygosity (Luo *et al.*, 2000) resulting in increased tumorigenicity (Goss *et al.*, 2002; Luo *et al.*, 2000). The increased genomic instability of the *Blm* mutant mouse and resulting loss of heterozygosity is being used to identify novel tumor suppressor genes in mice (discussed below) (Suzuki *et al.*, 2006). In addition to the mouse model of Bloom syndrome, many models have been developed targeting specific steps in the DNA damage response and DNA repair, allowing researchers to study how DNA instability/damage alters tumor phenotypes (e.g., see Wei *et al.* (2002) for models of DNA mismatch repair genes; see Hande (2004) for DNA damage signaling genes; see Barlow *et al.* (1996) and Westphal *et al.* (1997) for *Atm* mutant models).

In order to assay *in vivo* for genetic instability caused by toxins and carcinogens, many groups make use of the commercially available Big Blue® mouse and Muta™ Mouse animal models (Stratagene, La Jolla, CA). Both models harbor multiple copies of a mutational target within a recombinant phage that can be extracted from the genomic DNA of mouse tissues. For a review of animal mutation models, see Nohmi *et al.* (2000). In the Big Blue® Transgenic Mouse Mutagenesis Assay System, 30–40 copies of the *lacI* repressor gene have been stably integrated in tandem into the mouse genome (Kohler *et al.*, 1991). Big Blue® mice are either treated with mutagenic compounds or crossed with genetically engineered mouse models and used to determine changes in chromosomal stability by measuring mutagenesis of the *lacI* gene (Fig. 3). Mutation of *lacI* disables its ability to repress *lacZ* expression when the extracted phage is infected into host bacteria. *LacZ* is measured in bacteria by growing on substrate that turns blue in the presence of enzymatic activity of the product of the *LacZ* gene. The ratio of blue to colorless plaques is used as a measure of mutagenicity.

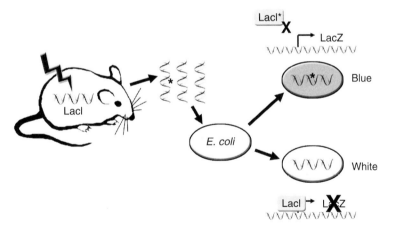

Fig. 3 Measuring mutagenesis levels using the Big Blue® mouse. Big Blue® mice have been engineered to carry many copies (30–40) of the bacterial *lacI* gene, encoding the LacI repressor protein. When Big Blue® mice are subjected to mutagens or crossed to mice with a mutator phenotype, the *lacI* gene is mutated. Genomic DNA is isolated from the mice and infected into *lacI*⁻ bacteria via lambda phage. Bacteria carrying unmutated *lacI* continue to repress the *lacZ* gene, whereas bacteria carrying mutant *lacI* expressed *lacZ* and form blue colonies on X-gal indicator plates. The ratio of blue to clear bacterial colonies is a measure of mutagenesis.

The Muta™Mouse uses a similar system, but relies on negative selection in measuring mutagenesis of *lacZ*, rather than *lacI* (Gossen *et al.*, 1989). While these systems rely on colorometric differentiation to identify mutations, addition of *cII* selection measures cell survival using bacteriophage lambda lytic and lysogenic multiplication cycles. Using the *cII* system in conjunction with the Big Blue® or Muta™Mouse can lead to better quantification of mutation frequencies, reduction of false positives, and reduced cost.

E. Discovery of Novel Tumor Genes

1. INSERTIONAL MUTAGENESIS

Forward genetic screens can identify cancer-causing genes by unbiased, whole genome mutation screening in mouse tumors. Chemical mutagenesis has been used to produce a high frequency of mutations in the mouse. Phenotypic screens of these mutant mice identify cancer-related phenotypes, but gene identification is very difficult. Alternatively, insertional mutagenesis allows for identification of mutation sites using the integrant as a molecular tag that marks the region of the genome important for tumorigenesis.

Retroviruses have been used as insertional mutagens to identify cancer-causing genes since it was determined that certain strains of mice develop leukemia while other strains develop mammary tumors due to ecotropic retroviral infection (Gross, 1978; van Lohuizen and Berns, 1990). The cancers in these mouse strains are not due to acute transforming retroviruses that express viral oncogenes such as *v-abl* or *v-myb* (reviewed in Lipsick and Wang, 1999; Shore *et al.*, 2002). Rather, these slow-transforming retroviruses, such as murine leukemia virus (MuLV) and mouse mammary tumor virus (MMTV), lead to transformation by acting as insertional mutagens that integrate into the genome and disrupt the normal expression of oncogenes and tumor suppressor genes.

Retroviral integration is regulated by the long terminal repeats (LTRs) that contain enhancer and promoter regions, as well as start and stop sites for transcription of retroviral genes. Retroviruses disrupt gene expression by four basic mechanisms, dependent on the viral location and orientation relative to the gene (Jonkers and Berns, 1996). Viral integrations can cause gene mutations by promoter insertion, disruption of RNA or protein destabilization motifs, viral enhancement of transcription, or premature termination of genes (Jonkers and Berns, 1996, reviewed in Uren *et al.*, 2005) (Fig. 4). For example, proviral insertion in the gene promoter region (Fig. 4A) places the gene under the control of the LTR promoter region and has been found to upregulate the oncogenes *LCK*, *N-Ras*, and *E2a* in MuLV-induced leukemia, as well the novel candidate oncogene *Evi-1*

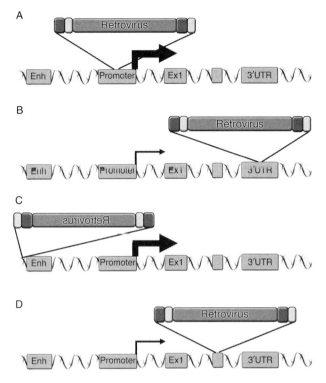

Fig. 4 Mechanisms of retroviral mutagenesis. Because of the presence of enhancer and promoter sequences in the LTR regions of the retrovirus, there are four basic mechanisms by which retroviruses disrupt gene function. (A) Retroviruses can insert into gene promoters, providing a stronger promoter signal from the LTR and increasing the level of gene expression. (B) Retroviruses can insert into the 3′UTR of a gene, changing mRNA stability thus altering the amount of protein translated. (C) Retroviruses can act as enhancers of gene transcription when inserted upstream, or downstream, of the gene. (D) Retroviruses can mutate genes by inserting into an exon and causing premature truncation of the protein.

(Morishita *et al.*, 1988, reviewed in Hirai *et al.*, 2001). Disruption of 3′ UTR sequences due to proviral integration causes activation of *N-myc* in MuLV-induced leukemia (van Lohuizen *et al.*, 1989) (Fig. 4B). *Wnt1* and *Fgf3* are commonly overexpressed in MMTV-induced mammary tumors, due to activation by enhancer sequences of the integrated MMTV provirus, often at a distance up to 25 kb 5′ or 3′ of the gene (Fig. 4C). Although tumor suppressor inactivation is less common, mice infected with Friend leukemia virus have inactivating viral integrations within the *p53* gene (Ben David *et al.*, 1988; Mowat *et al.*, 1985), and *Nf1* is often inactivated in MuLV leukemia (Cho *et al.*, 1995; Largaespada *et al.*, 1995) (Fig. 4D). By utilizing the virus as a molecular marker, many groups have cloned sites of integration to identify genes involved in specific types of cancer. High-throughput methods of cloning using PCR-based methods have allowed the identification of a large number of sites, compiled in the Retroviral Tagged Cancer Gene Database (http://rtcgd.ncifcrf.gov/) (Akagi et al., 2004). Common sites of integration, sites that are disrupted in multiple independently derived tumors, are more likely to play a causal role in tumor progression pathways.

One drawback of retroviral insertion is that the majority of tumors are caused by gene activation of proto-oncogenes, suggesting that this model is inefficient at identifying tumor suppressor genes. However, as technologies improve and greater numbers of insertion sites are rapidly and completely identified (reviewed in Kool and Berns, 2009), the increased number of insertions sites per tumor will improve the frequency of identifying tumor suppressor genes (Uren et al., 2008), especially those with haploinsufficient tumor suppressor functions. Additionally, in order to increase loss of function by viral insertion, screening can be combined with the *Blm* mutant mouse (Suzuki et al., 2006) that has a higher rate of mitotic recombination and increased rate of loss of heterozygosity (discussed above) (Luo et al., 2000). When the *Blm* mutation was crossed onto mouse strains prone to retroviral insertion-mediated B-cell lymphomas, cancer latency was reduced, and loss of heterozygosity was found in multiple known and novel tumor suppressor genes (Suzuki et al., 2006). The increased efficiency of retroviral oncogene activation compared to tumor suppressor gene inactivation may be due to the preference of retroviruses to integrate in the 5′ end of genes (Johnson et al., 2001). This promoter region preference may also indicate that only specific regions of the genome are accessible to viral integration, suggesting that whole genome coverage by viral insertional mutagenesis may be impossible. Although retroviral integration has been useful in identifying candidate cancer genes, these models generally only develop hematopoietic and mammary tumors.

2. MUTAGENESIS BY TRANSPOSONS

In an attempt to utilize somatic mutagenesis beyond hematopoietic and mammary tumorigenesis, and perhaps to more fully cover the genome with mutagenic events, DNA transposons are now being used as a mobile genetic tag in the mouse. Transposable elements have been used in multiple organisms for years, but the first active transposon in the mouse is *Sleeping Beauty*, a member of the *Tc1/mariner* family, which was reconstructed by directed mutagenesis from a dormant element (Ivics et al., 1997). More recently, *piggyBac* has been demonstrated to be an efficient transposable element in mammalian cells *in vitro* and *in vivo* (Ding et al., 2005; Wilson et al., 2007). Both systems consist of two elements: the transposon (the mobile DNA

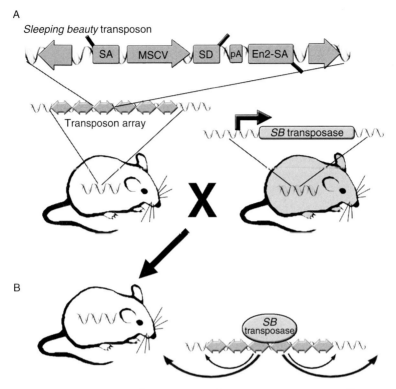

Fig. 5 Transposon insertional mutagenesis with *Sleeping Beauty*. Mice genetically engineered to carry arrays of transposon insertions are crossed to mice genetically engineered to express the transposase. (A) The *Sleeping Beauty* T2/Onc2 transposon structure is shown (Dupuy *et al.*, 2005) in which the transposon carries two splice acceptors (SA) in opposite directions, a bidirectional poly(A) tail (pA), and the MCSV LTR (MCSV) with a splice donor (SD) that can act as a promoter/enhancer of gene expression. These sequences allow for both gain-of-function and loss-of-function mutations when inserted into a gene locus. The transposon is carried by transgenic mice in arrays of 150–350 transposon copies. Transposon-carrying mice are crossed to mice expressing the *Sleeping Beauty* (*SB*) transposase from a variety of different promoters to induce mutagenesis in different tissues. (B) The progeny of transposon and transposase mice express the tranposase and cause both intrachromosomal and interchromosomal hopping of the transposon to mutate the genome.

sequence) and transposase (the enzyme that mobilizes it) (Fig. 5). The T2/onc (Collier *et al.*, 2005) and T2/onc2 (Collier *et al.*, 2005; Dupuy *et al.*, 2005) *Sleeping Beauty* transposons are currently the most commonly used for somatic mutagenesis in cancer gene identification experiments. These transposons contain splice acceptors followed by polyadenylation sequences in both orientations in order to generate loss-of-function mutations, as well as retroviral LTR enhancer/promoter elements to drive overexpression of nearby genes. Two groups have shown the ability of *Sleeping Beauty* to mobilize in somatic cells, leading to tumor formation (Collier *et al.*, 2005; Dupuy *et al.*, 2005). Although the transposons are similar, the transgenic lines vary greatly in copy number. The lines contain multicopy arrays of the transposons at the initial site of insertion; T2/onc lines contain approximately 25 copies (Collier *et al.*, 2005), whereas T2/onc2 transgenic lines contain 150–350 copies of the transposon (Dupuy *et al.*, 2005). The differences in copy number are believed to contribute to the phenotype seen in the mice (Collier *et al.*, 2005; Dupuy *et al.*, 2005). The T2/onc2 lines cause tumorigenesis in an otherwise wild-type background; however, these mice develop aggressive hematopoietic tumors that kill the mice prior to the development of any solid tumors (Dupuy *et al.*, 2005). Although the T2/onc line does not cause tumorigenesis in wild-type mice, a wide variety of tumors form when crossed onto a $p19^{Arf}$ deficient background, including osteosarcomas, soft tissue sarcomas, lymphomas, malignant meningiomas, myeloid leukemia, and pulmonary adenocarcinoma (Collier *et al.*, 2005), similar to the tumor spectrum seen in $p19^{Arf-/-}$ mice on the C57BL/6 genetic background (Kamijo *et al.*, 1999). Crossing different transposon and transposase transgenic lines allowed the system to be refined to generate high-penetrance hematopoeitic tumors without secondary effects of embryonic lethality (Collier *et al.*, 2009). Furthermore, by combining a Cre-inducible *Sleeping Beauty* transposase with a third generation *Sleeping Beauty* transposon line (T2/Onc3), Dupuy *et al.* (2009) were able to successfully control the extent of hematopoeitic tumors to generate a wide variety of mouse solid tumors by insertional mutagenesis. Because *Sleeping Beauty* shows preferences for local integration near the original transposon site, the system may have some limitations for saturating the genome with mutations. However, the differences in accessibility of the genome to mutation by DNA transposons and mouse retroviruses may allow saturating mutagenesis through a combination of the two techniques, particularly as mouse panels are developed with DNA transposon arrays integrated systematically onto the different mouse chromosomes.

In order to more accurately model specific tumor types, a conditional *Sleeping Beauty* system has been developed by knocking a *lox-STOP-lox SB11* transposase allele into the *Rosa26* locus (Dupuy *et al.*, 2009; Keng *et al.*, 2009). Expression of the *Sleeping Beauty* transposase can therefore be activated using a tissue-specific Cre to drive transposition in the tissue and developmental stage of interest. A model of hepatocellular carcinomas was developed by crossing the *Sleeping Beauty* line to a hepatocyte-specific albumin-Cre (Keng *et al.*, 2009), and a colorectal cancer mouse model was developed by crossing to a Villin-Cre transgenic mouse line that expresses Cre only in the epithelial cells of the gastrointestinal tract (Starr *et al.*, 2009).

Analysis of the insertion sites in both models identified multiple common insertion sites near genes known to be mutated in hepatocellular carcinoma and colorectal cancer, as well as genes not previously associated with these cancers (Keng *et al.*, 2009; Starr *et al.*, 2009). These models demonstrate the ability of transposons to identify relevant cancer-causing genes, in a tissue-specific manner in solid tumors.

3. GENETIC MODIFIER SCREENS

Many of the genetic mutations that have been identified as cancer causing are rare mutations that have highly penetrant effects on tumor susceptibility. Studies have shown that more common variants with subtle effects, such as polymorphisms seen in different mouse strains, also contribute to tumorigenesis. Different inbred strains show large differences in susceptibility to different tumor types, confirming the ability of low-penetrance polymorphisms to modify cancer risk (reviewed in Dragani, 2003). Similarly, the analysis of multiple human studies has shown that the majority of cancers occur in individuals with a genetic predisposition (reviewed in Demant, 2003). These studies suggest that the combined contributions of low-penetrance genes have a larger effect on cancer morbidity and mortality than the rare mutations that have been identified (reviewed in Demant, 2003). Modifier genes have been seen to affect a variety of tumorigenic phenotypes including overall resistance, metastatic ability, angiogenic potential, multiplicity, size, and survival.

As with quantitative trait loci (QTL) screens in mouse models of many human diseases (Hunter and Crawford, 2008; Peters *et al.*, 2007), modifier screens identify cancer susceptibility loci in the mouse by taking advantage of the known phenotypic differences between strains. The sequencing of the genome and the identification of genetic markers, such as simple sequence length polymorphisms, microsatellite markers, and single nucleotide polymorphisms (SNPs), have increased the ability to identify these low-penetrance modifier genes. Generally, modifier screening is based on crossing two parental strains, one resistant and one highly susceptible to a specific tumor, then backcrossing or intercrossing the F1 progeny to identify those genetic regions which affect the tumorigenic phenotype in a statistically significant way (Fig. 6A).

Even with the increase of genetic markers, many modifier loci have been found but few specific genes have been identified due to the difficulty of narrowing down a genetic locus to a single gene via positional cloning. Recent advances are being utilized to increase the ability to identify the specific genes that underlie QTL, including cancer modifiers. These include both genomic and animal resources, as well as new techniques and analytical tools (reviewed in Flint *et al.*, 2005). Some current studies have taken

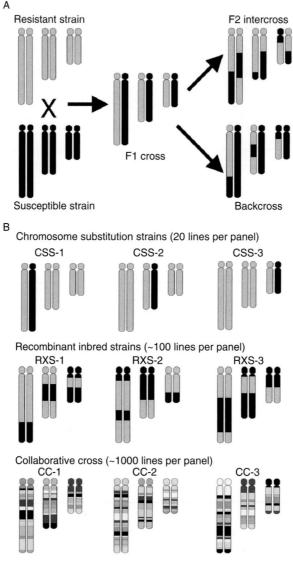

Fig. 6 Genomic tools for mapping cancer modifiers in mouse. (A) Classic mapping approaches have used two generations of breeding to generate recombinants in the genome, either through F2 intercrosses or backcrosses. Individual mice are then genotyped at polymorphic loci between the resistant and susceptible strain to correlate the genotype to the modifier phenotype. (B) Reference strain panels, such as chromosome substitution strains, recombinant inbred lines, and the Collaborative Cross, provide stable, genotyped recombinants and can be compared for their cancer phenotype to directly map modifiers, or can be crossed to genetically engineered mouse models of cancer to map modifiers in the first generation.

advantage of recombinant inbred strains (Fig. 6B), panels generated from an initial intercross of two inbred strains, followed by repeated brother–sister matings, which result in each recombinant inbred strain containing unique combinations of the progenitor genomes. These strains have been a useful resource for mapping by contrasting phenotypes and genotypes between individual recombinant inbred strains, but their use in establishing linkage to low penetrance genes is limited by both the small set size of available recombinant inbred panels, and the small number of different recombinant inbred panels that have been generated (Demant, 2003; Flint et al., 2005). A more recent development has been the use of chromosome substitution strains (Fig. 6B), in which one chromosome comes from a different strain background than the other 20. This contributes to increased QTL detection by reducing potential epistatic interactions between the modifier of interest and modifiers on other chromosomes, and allowing initial studies of modifier function (Flint et al., 2005; Nadeau et al., 2000). Additionally, backcrossing the chromosome substitution strains can further increase resolution of modifier mapping (Hunter and Crawford, 2008).

A current proposal to further ease the difficulty in cloning modifier genes is an expansion of the idea of the recombinant inbred strain. This Collaborative Cross resource, suggested by the Complex Trait Consortium (Churchill et al., 2004) is currently being generated through the collaboration of multiple groups (Chesler et al., 2008; Iraqi et al., 2008; Morahan et al., 2008). The Collaborative Cross is a panel of an estimated 1000 recombinant inbred lines being generated, following a specific breeding scheme, from eight divergent founder strains A/J, C57BL/6J, 129S1/SvImJ, NOD/LtJ, NZO/HiLtJ, CAST/Ei, PWK/PhJ, and WSB/EiJ (Chesler et al., 2008; Iraqi et al., 2008; Morahan et al., 2008). Each new recombinant inbred line will be genotyped initially, resulting in a large number of genetically defined strains to be used in comparative studies (Fig. 6B) (Churchill et al., 2004). The Collaborative Cross, when fully inbred, will have a QTL mapping resolution of approximately 1 MB (Broman, 2005; Valdar et al., 2006), much higher than is currently possible. Because many inbred strains are closely related and only differ in a subset of their genome, comparison of any two inbred strains (as is the case with backcrossing or intercrossing (Fig. 6A), and chromosome substitution strains or recombinant inbred strains (Fig. 6B)) leads to "blind spots" in the genome where the strains are identical by descent. The Collaborative Cross takes advantage of more diverse strains, increasing the amount of genetic diversity between the strains, and thus the extent of the genome available to be screened (Churchill et al., 2004). In conjunction with the Collaborative Cross, the Jackson Laboratory is generating the Diversity Outbred Mouse Population to further improve resources for gene mapping. The diversity outcross is a heterogeneous stock population that will be maintained as an outbred stock, approximating the heterogeneous human population. Diversity Outbred mice will be generated from 160 breeding lines of the Collaborative Cross, allowing the early recombination events to be maintained while avoiding inbreeding (Lambert, 2009).

As mentioned above, although many modifier loci have been mapped, few genes have been cloned (reviewed in Flint *et al.*, 2005), with the number involved in cancer being even smaller. The first locus to be mapped as a cancer modifier in the mouse was the *Mom1* locus on chromosome 4 (Dietrich *et al.*, 1993), found to modify the *Apc* gene mutation, *Min* (*Apc^{min}*), an inducer of intestinal cancer (Moser *et al.*, 1992). Additional studies showed that *Mom1* modified not only intestinal tumor number, the phenotype for which it was identified, but also tumor size (Gould *et al.*, 1996). Further analysis identified secretory phospholipase A2 (*Plag2g2a*) as a candidate gene for the *Mom1* locus (MacPhee *et al.*, 1995); *Plag2g2a* was later shown to functionally modify the *Apc* mutation (Cormier *et al.*, 1997). Interestingly, studies in humans suggested that *PLA2G2A* does not play a major modifying role in human colorectal cancers, due to a lack of functional polymorphic differences in the human gene (Riggins *et al.*, 1995; Tomlinson *et al.*, 1996). Although this highlights the differences between mouse and human, additional *Modifier of Min* (*Mom*) genes have been mapped, and the genes are currently being identified and tested for their role in human cancer.

In contrast, the gene *Sipa1*, originally identified as the gene underlying the metastasis modifier loci *Mtes1*, is associated with breast cancer tumorigenesis in both mouse and humans. The MMTV-polyoma middle T (MMTV-PyMT) transgenic mouse model develops mammary tumors as well as extensive metastases, with metastatic growth being influenced by strain background (Lifsted *et al.*, 1998). QTL mapping identified *Mtes1* as a metastasis modifier locus on mouse chromosome 19 (Hunter *et al.*, 2001), and the region was further narrowed down by taking advantage of haplotype analysis (Park *et al.*, 2003b). A combination of sequence and functional analysis originally identified *Sipa1* as a polymorphic candidate for the *Mtes1* locus (Park *et al.*, 2005). SNP analysis of human breast tumors found that germline polymorphisms of the *Sipa1* gene are associated with metastatic potential and poor prognosis (Crawford *et al.*, 2006), demonstrating that genes identified by modifier screens in the mouse can play a role in human cancer.

Another example of the complexities of modifier interactions in cancer is the *nerve sheath tumor resistance* locus (*Nstr1*), a modifier locus on mouse chromosome 19 found to affect incidence of peripheral nerve sheath tumors in an $Nf1^{-/+};p53^{-/+}$ cis mouse model of neurofibromatosis type 1 (Reilly *et al.*, 2006). Interestingly, the *Nstr1*-modifying effect is dependent on epigenetic factors on other chromosomes, with susceptibility differences only being seen if the *Nf1;p53* mutant chromosome 11 was passed on to the progeny from the father. More recent studies used a chromosome substitution strain with a C57BL/6J background and A/J chromosome 19 to show that a modifier on chromosome 19 affects both the incidence and latency of peripheral nerve sheath tumors, as well as astrocytoma, another tumor type seen in the $Nf1^{-/+}p53^{-/+}$ cis mouse model but not previously believed to be modified by a locus on chromosome 19. Additionally, the effects of the A/J chromosome 19 vary based not only on inheritance of the $Nf1;p53$ mutant chromosome as seen previously but also on the sex of the progeny (Walrath *et al.*, 2009). These findings demonstrate not only the complex interactions between genetic background and tumorigenesis but also stress the importance of interpreting the role of genes in tumorigenesis in the context of the genetic background being studied.

III. CANCER PARADIGMS AND LESSONS FROM THE MOUSE

Although much has been learned over the past three decades about the cellular and genetic changes that occur within the cancer cell, there is a growing appreciation that the environmental context in which the cancer cell develops is critical for how cancer forms. This highlights the importance of performing cancer cell experiments in more physiologically relevant contexts to recapitulate the environments experienced by the cancer cell in the patient. Mouse model systems can provide experimentally tractable systems to study the role of development and environment in cancer. We describe here ongoing research into the distinction between different cell types within a tumor, with a focus on the cancer stem cell (CSC), the role of the microenvironment, with special emphasis on the vasculature and immune system, and metastasis, in which cancer cells leave their current environment and adapt to survive at distant sites from the tumor. We discuss how mouse models can help to better understand these non-cell-autonomous aspects of cancer.

A. The Cancer Stem Cell and Initiating Cell

In order to eradicate tumors and prevent recurrence, it is important to understand tumor growth dynamics and the cellular subpopulations driving tumor growth. A comparison of the current hypotheses of tumor growth mechanisms is reviewed by Adams and Strasser (2008). According to the **CSC model** (Fig. 7), tumor growth is driven by a rare population of CSCs that maintains the ability to self-renew and gives rise to more differentiated cell types (reviewed in Tan *et al.*, 2006). This theory is based on evidence from mouse and human studies showing that malignant tumors, such as acute myeloid leukemia, contain a subpopulation of cells that have stem

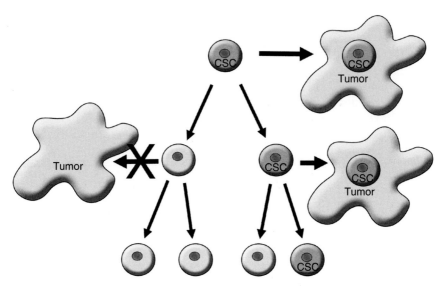

Fig. 7 The cancer stem cell model. In this model, tumors contain rare populations of cells with the ability to repopulate the tumor (cancer stem cells; CSC). These CSCs divide asymmetrically to give rise to more differentiated cells (light gray) and additional CSCs (dark gray). Only the CSCs can give rise to tumors. More differentiated cells can proliferate, but are not sufficient to repopulate the tumor.

cell-like self-renewal and differentiation characteristics, express stem cell markers, and can engraft into immune-compromised NOD-SCID mice. The **clonal evolution model** postulates that tumor growth can be maintained by a substantial portion of the tumor cells that are proliferating rapidly and that may be at various stages of the differentiation pathway rather than a specific CSC. A third mixed model provides that a clone with CSC properties arises late as a dominant subclone after selection within the growing tumor. Given the broad diversity of cancer, it may be impossible to generalize across all tumor types and it is likely that multiple models accurately describe tumors from different origins (Gupta *et al.*, 2009). Therefore, the predominant growth model for each tumor type must be determined experimentally.

In contrast to the CSCs that can maintain tumor growth within established tumors, the tumor-initiating cell is the cell that first undergoes mutation in the progression to cancer. There is much ongoing debate as to whether tumors must initiate in a normal stem cell compartment, in a cell with ongoing proliferative capacity, or whether normal, differentiated cells arrested in the G_0 phase of the cell cycle can give rise to tumors. Identification of the tumor-initiating cell will provide insight into the process of tumorigenesis that may lead to new ideas in cancer prevention and treatment. While the idea that CSCs are present in many tumors is widely accepted, how and where they come from is still under debate. One possibility is that CSCs arise from stem/progenitor cells that have lost the ability to senesce or differentiate properly, as has been shown for acute myeloid leukemia and promyelocytic leukemia. Another possibility is that more differentiated cells have acquired stem cell-like characteristics (Hambardzumyan et al., 2008), as has been suggested by work on astroglia (Moon et al., 2008; Steindler and Laywell, 2003), oligodendroglia (Grinspan et al., 1996), myocytes (Odelberg, 2002), adult mouse fibroblasts, and human somatic cells (Takahashi and Yamanaka, 2006). Whether the tumor-initiating cell that gives rise to the CSC comes from a stem/progenitor cell gone awry or a more differentiated cell with acquired stemness, the identity of the tumor-initiating cell is likely to be a major determinant of tumor characteristics.

Conditional knockout mice with tissue-specific Cre expression have been used to identify stem/progenitor cell types that can serve as tumor-initiating cells. Conditional loss of *Nf1* in GFAP$^+$ progenitor cells of the subventricular zone is sufficient to drive astrocytoma formation in mice, suggesting that stem/progenitor cells can be tumor-initiating cells for astrocytoma (Zhu et al., 2005). Conditional loss of *Nf1* in fetal neural crest stem/progenitor cells of the Schwann cell lineage leads to development of abnormal differentiated Schwann cells, which then serve as neurofibroma tumor-initiating cells (Zheng et al., 2008). This suggests that the neural crest stem/progenitor cell is not the direct tumor-initiating cell, but that deregulation at the stem/progenitor level can develop into an abnormal cell type that initiates tumors. Although conditional genetically engineered mouse models facilitate elegant studies that indicate which cell types are sufficient to initiate tumorigenesis, induction of genetic changes in specific cell types does not test whether tumorigenesis must initiate in the tested cell population, or whether alternate pathways to tumorigenesis are possible. Thus, the true identity of the tumor-initiating cell in spontaneous tumors is still unclear for most tumor types. The development of additional mouse models to identify tumor-initiating cells in spontaneous tumors would help address this question. Unfortunately, the identification of tumor-initiating cells has been largely dependent on the definition of stem/progenitor cell lineages using markers of cell differentiation. These markers help define the cells being studied and are used to specifically drive Cre expression in cell populations to test hypotheses of tumor initiation. Whereas cell lineage and characteristic markers of the different stages of differentiation are well defined in the hematopoietic system, other organ systems are not as clearly understood, making it difficult to study in which cells tumors arise. Furthermore, as more stem/progenitor cell markers are identified and more differentiation steps are defined, the developmental steps from a stem cell to more committed progenitor cells and ultimately differentiated cell types may expand. Therefore, the cell identities capable of initiating tumors are likely to be refined as more stem/progenitor cell types along differentiation pathways are identified.

The progression of stem cells from totipotent to pluripotent to developmentally restricted progenitors is controlled by changes in expression of transcription factors through chromatin remodeling and modification, generally referred to as epigenetic changes (Niwa, 2007; see reviews by Zardo *et al.* (2008) and Jones and Baylin (2007) for more information). Alterations in epigenetic control, such as changes in chromatin regulators and chromatin modifications, may alter the expression of lineage-specific genes in stem cells during development, thereby inhibiting proper differentiation into more committed progenitors or differentiated cells and leading to expansion of cells with CSC potential. For example, overexpression of the chromatin regulator, *Bmi1*, in hematopoietic stem cells of transgenic mice results in increased stem cell self-renewal and leukemia (van Lohuizen *et al.*, 1991), while loss of *Bmi1* blocks leukemic proliferation and transplantation (Lessard and Sauvageau, 2003; Park *et al.*, 2003a).

Mouse models have been used to study the importance of regulation of epigenetic determinants, such as DNA methylation. The control of gene transcription and epigenetics is tightly regulated by DNA methyltransferases (DNMTs). Changes in DNA methylation profiles reflect changes in epigenetic states and cell fate decisions, including stemness (Bibikova *et al.*, 2006) and tumorigenesis (Lee *et al.*, 2008; Martinez *et al.*, 2009), as well as playing a critical role in genome stability (Eden *et al.*, 2003; Gaudet *et al.*, 2003). Studies that compare the role of *de novo* DNMTs, *Dnmt3a* and *Dnmt3b* (Okano *et al.*, 1999), to the maintenance DNMT, *Dmnt1* (Hirasawa *et al.*, 2008; Lei *et al.*, 1996), have helped to delineate the role of DNMTs in development and cell differentiation. Embryonic stem cells lacking DNMTs are severely hypomethylated and fail to differentiate. In stem cells with reduced levels of methylation, *Dnmt1* is critical for stem cell differentiation, whereas embryonic stem cells in which the *de novo* DNMTs are mutated can still differentiate, despite similar levels of DNA methylation in the two situations. These data emphasize that the regulation of stem cell epigenetics and differentiation capacity is dependent not on the methylation levels *per se*, but on the mechanism by which methylation is maintained (Jackson *et al.*, 2004). Given that tumors are often globally hypomethylated and locally hypermethylated and that upregulation of *DNMT1* correlates with poorer prognosis (Kanai, 2008), it will be important to fully understand the relationship between normal epigenetic control of stem cells and the role of epigenetic changes in cancer, both in terms of gene expression changes and genome stability. Genetically engineered mice with altered expression and/or function of epigenetic regulators, such as *Bmi1* and *Dnmt*, are powerful tools to study the epigenetics of stem cells and how deregulation of stem cell epigenetics can lead to cancer.

B. The Tumor Microenvironment

Cancer research has long focused on the mutated neoplastic cells and as a result early mouse models were designed to identify the effects of cell-autonomous mutations on tumorigenesis. Over the past decade, tumors have been found to consist of many nonneoplastic cell types, including fibroblasts, endothelial cells that form the vasculature and lymphatic system, and innate and adaptive immune cells, all of which interact to influence tumorigenesis (reviewed in Weinberg, 2008). There is some debate over whether stromal cells evolve and mutate to make them more capable of supporting tumor growth. Recent studies indicate that large DNA alterations are not present in fibroblasts, epithelial cells, or other neighboring stromal cells of human breast and ovarian cancers (Allinen et al., 2004; Qiu et al., 2008, reviewed in Weinberg, 2008), although expression changes occur suggesting that epigenetic changes may play a role in evolving tumor stroma. In contrast, studies in mouse models support the hypothesis that mutations in the microenvironment are important for tumorigenesis (Hill et al., 2005; Zhu et al., 2002). Additional experiments are needed to determine the role of mutations outside the neoplastic cells, as well as to determine how epigenetic changes in stroma contribute to tumorigenesis. Meanwhile, research suggests that the interaction between a tumor and the microenvironment causes a disruption in normal tissue homeostasis, resulting in an environment more conducive to tumorigenesis (Fig. 8).

1. INFLAMMATION

Chronic inflammation is one common mode of microenvironment disruption that contributes to a tumorigenic phenotype and has been found to affect tumor development, both in the clinic and in mouse models of tumorigenesis (reviewed in de Visser and Coussens, 2006; Tlsty and Coussens, 2006). Two different models of colon cancer, the *T-cell-receptor-β/Trp53* double knockout and the *Rag2/TGFβ*-deficient double knockout, do not develop tumors when maintained in germ-free housing, demonstrating a requirement for inflammation in tumorigenesis (Engle et al., 2002; Kado et al., 2001). Inflammation has also been implicated in a transgenic model of epithelial cancer in which the early genes of human papillomavirus type 16 are expressed under control of the human keratin 14 promoter (K14-HPV16) (Arbeit et al., 1994; Coussens et al., 1996, reviewed in de Visser and Coussens, 2006; de Visser et al., 2006). In this model, the premalignant skin is infiltrated by innate

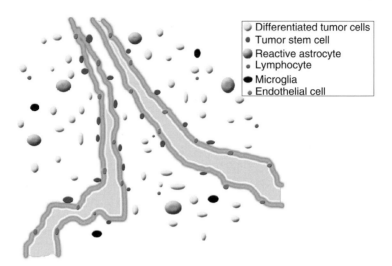

Fig. 8 The complexity of the tumor microenvironment. This cartoon shows an example of the tumor cell types making up the tumor microenvironment, based on the current understanding of brain tumors. In addition to differentiated tumor cells that make up the bulk of the tumor, tumor stem cells reside in a specific niche along blood vessels. Normal cells, such as reactive astrocytes, can react to the changes in the local microenvironment to become altered in morphology or gene expression. Inflammatory cells, such as lymphocytes and microglia, invade the region of the tumor and can further change the local microenvironment.

immune cells, primarily mast cells and granulocytes (Coussens *et al.*, 1999; de Visser *et al.*, 2005). In a mast cell-deficient background, the K14-HPV16 mice show attenuated tumor development, confirming the functional contribution of the immune cells to cancer development (Coussens *et al.*, 1999). Interestingly, a similar model in which HPV16 infection occurs in the cervical epithelium also shows an influx of immune cells, but in contrast to the skin model the immune cells in the cervix are infiltrating macrophages (Giraudo *et al.*, 2004) rather than mast cells. This demonstrates the specificity of changes in the microenvironment in different tumor types, even when initiated by the same oncogene. The importance of mast cells in tumor initiation has also been illustrated in a study of neurofibroma development. Although Schwann cells are the tumor-initiating cells in an *Nf1* conditional knockout, $Nf1^{+/-}$ mast cells are required for the development of neurofibromas, and infiltrate the area around the peripheral nerve prior to tumorigenesis (Yang *et al.*, 2008; Zhu *et al.*, 2002).

Inflammation has been found to affect not only premalignant progression and tumor initiation but also tumor growth and metastasis. In the polyoma-middle-T-antigen (PyMT) transgenic mouse model of mammary tumors,

PyMT mice lacking the cytokine CSF-1 had reduced macrophage recruitment to neoplastic tissue, but early neoplastic development was unaffected (Lin et al., 2001). However, lack of CSF-1 significantly delayed the development of invasive carcinoma, and pulmonary metastases were reduced. This metastatic ability was restored with transgenic CSF-1 expression in the mammary epithelium (Lin et al., 2001).

Inflammatory cells modulate tumor development by both direct and indirect effects on neoplastic cells. Cytokines and chemokines, as well as other inflammatory cell signaling molecules, promote tumor progression by several mechanisms (reviewed in Kundu and Surh, 2008). Two independent mouse models show that the signaling molecule nuclear factor κB (NF-κB), a proinflammatory transcription factor, may provide a link between inflammatory cells and signal pathway activation within neoplastic cells (reviewed in Balkwill and Coussens, 2004). In the *Mdr2*-knockout mouse model of inflammation-associated hepatocellular carcinoma, Pikarsky et al. (2004) demonstrate that hepatocyte NF-κB activation is controlled by inflammatory cell upregulation of tumor-necrosis factor α (TNF-α). In a colitis-associated colorectal cancer, IKKβ, an upstream positive regulator of NF-κB, has been shown to play a role in both the tumor cell and the inflammatory cell. When IKKβ is deleted in colonic enterocytes, the putative tumor-initiating cell, tumor incidence is decreased without affecting inflammation pathways. Myeloid-specific IKKβ deletion decreases tumor size through decreased production of cancer-promoting factors and a reduction in enterocyte proliferation (Greten et al., 2004). In combination, these studies suggest that inflammatory cells directly impact neoplastic cells by promoting proliferation through secretion of growth factors into the tumor microenvironment. Immune cells are also known to have an indirect role on tumor growth by impacting the cellular microenvironment through tissue remodeling and angiogenesis. Mouse models of inflammation and immunomodulation are reviewed by de Visser et al. (de Visser and Coussens, 2006; de Visser et al., 2006).

It is clear that inflammation can have a tumor-promoting effect through the stimulating effect of cytokines on tumor cells and the ability of inflammatory cells to remodel the microenvironment. However, it should be noted that certain types of innate immune cells, particularly type 1 natural killer T cells, have an antitumorigenic response, as do some cytokines (reviewed in Fujii, 2008; Terabe and Berzofsky, 2007, 2008). Genetically engineered mouse models lacking certain cell types of the immune system, or certain cytokines, have been critical in understanding the role of immune cells on tumor immunity. For example, the mice mutant for Vα14Jα18 (Jα18 mice) lack type 1 natural killer T-cells, whereas CD1d mutant mice lack both type 1 and type 2 natural killer cells. Comparison of tumor growth in these two mouse models has led to the understanding that type 1 natural killer T cells promote tumor immunesurveillance, whereas type 2 natural killer T cells suppress this immunosurveillance (reviewed in Terabe and Berzofsky, 2007). This suggests that inflammation can have both a positive and negative effect on tumor growth, stressing the importance of understanding the balance between normal and abnormal regulation of the immune response.

2. ANGIOGENESIS

It was originally proposed by Folkman (1972) that cancers need to be vascularized to grow beyond a certain size and become malignant, and this theory has been supported over the years. The induction of vasculature growth to provide sufficient nutrients to the tumor has been termed the angiogenic switch, and occurs at varying stages of tumor progression, dependent on tumor type and environment. It was originally believed that tumor cells themselves were driving tumor angiogenesis, but recent data suggests that inflammatory cells play an important role by impacting tissue remodeling and activation of angiogenesis. The angiogenic switch is regulated by the balance between pro- and antiangiogenic factors. Normal angiogenesis is tightly regulated, whereas tumors lose appropriate control, such as failure of endothelial cells to become quiescent, allowing for constant growth of tumor blood vessels. Two classic mouse models of tumor angiogenesis are the K14-HPV16 mouse, described above, and the RIPTag mouse model of pancreatic islet carcinoma (Hager and Hanahan, 1999; Parangi et al., 1996). Neovascularization occurs early in dysplasia in these models, and is required for tumor formation. The tumors develop in temporally and histologically distinct stages that can be characterized by angiogenic status: normal cells, hyperplasia, angiogenic dysplasia, and last, highly vascularized invasive tumors.

The angiogenic factor vascular endothelial growth factor A (VEGF-A) has been confirmed to play a role in tumor development in multiple mouse models, including the RIPTag model mentioned above (reviewed in Crawford and Ferrara, 2009). In the APC^{min} mouse discussed above, studies suggested that VEGF was upregulated in adenoma epithelial and stromal cells. Treatment with the VEGF inhibitor, Mab G6-31, was found to arrest tumor growth but not prevent incidence, supporting the hypothesis that upregulated VEGF was contributing to tumor progression, but may not play a role in initiation (Korsisaari et al., 2007). Interestingly, the effectiveness of VEGF inhibition was found to be time dependent, such that when mice are treated earlier, the tumor number is reduced (Goodlad et al., 2006), whereas at later times only reduction in tumor size is seen (Goodlad et al., 2006; Korsisaari et al., 2007). In the MEN1 mouse model of multiple endocrine neoplasia (Crabtree et al., 2001), anti-VEGF treatment has been found to reduce vascular density and tumor size in both pituitary and pancreatic tumors, suggesting blocking of VEGF may be effective in treating multiple tumors within the same individual (Korsisaari *et al.*, 2008). In the RIP-Tag2 model of pancreatic islet carcinogenesis, angiogenesis is involved in the progression of tumorigenesis. Although VEGF-A expression is not significantly upregulated in tumors, VEGF deletion in the B cells of the Rip-Tag mice reduces angiogenesis and attenuates tumor growth, supporting the necessity for VEGF in tumor growth. Matrix metalloproteinase 9 (MMP-9) promotes angiogenesis in the RIP-Tag model and it also increases the bioavailability of VEGF, providing a potential explanation for VEGF dependence without mRNA upregulation (Bergers *et al.*, 2000). Certain antiangiogenic therapies, such as the therapeutic antibody bevaciumab, specifically inhibit only human VEGF (Yu *et al.*, 2008), making it difficult to test the effects of these drugs in mouse models. A knockin mouse has been developed that replaces 10 of the 19 differing amino acids between mouse and human VEGF, under the control of the endogenous murine promoter, minimizing developmental abnormalities. The biological activity of "humanized" VEGF is similar to human VEGF, as is its interaction with VEGF-blocking antibodies (Gerber *et al.*, 2007) (reviewed in Crawford and Ferrara, 2009).

The many cell types and cellular interactions in the microenvironment have made it difficult to determine which changes within the tumor are truly causal. Mouse models with activation and deactivation of specific oncogenic mutations *in vivo* have been created to follow the chain of tumorigenic events that occurs following oncogene mutation. In a Myc model of pancreatic cancer, MycER^{TAM}, a Myc fusion protein with a modified estrogen receptor-binding domain is activated in the presence of 4-hydroxytamoxifen (Lawlor *et al.*, 2006). Removing 4-hydroxytamoxifen stops Myc activity and allows analysis of both the Myc requirement and the timeline of events to tumorigenesis. Activation of Myc in islet cells triggers apoptosis, leading to islet involution. However, when apoptosis is blocked by coexpression of Bcl-x$_L$, Myc activation triggers pancreatic β cell expansion, leading to highly vascularized carcinoma (Pelengaris *et al.*, 2002). Myc overexpression in the islet cell compartments leads to the activation of angiogenic factors, such as VEGFA, that recruit endothelial cells and drive angiogenesis. Additional studies with this model suggest that VEGFA is released from the extracellular matrix via MMPs, such as MMP-9 (Shchors *et al.*, 2006) (reviewed in Shchors and Evan, 2007), and that secretion of interleukin-1β (IL-1β) from the β islet cells (Maedler *et al.*, 2002) is also important for the release of VEGFA and tumor angiogenesis.

There are exceptions to the angiogenic requirement for tumor growth. For example, astrocytomas are initially able to co-opt the blood supply from normal brain blood vessels without initiation of angiogenesis. Typically, astrocytomas develop along blood vessels without a tumor capsule, allowing growth and causing invasion into the brain. At higher grades, astrocytomas and glioblastomas become hypoxic and necrotic due to increased proliferation, tumor size, and ANG2-activated vessel regression (D'Angelo et al., 2000). Hypoxia then induces VEGF, promoting vascular remodeling and vessel sprouting, defining the progression from grade III astrocytoma to grade IV glioblastoma (Bergers and Benjamin, 2003; Theurillat et al., 1999). In this case where tumor cells are highly invasive from the early stages of tumorigenesis, angiogenesis is only required at late stages of progression when the tumor bulk overwhelms the normal vasculature system.

Different tumors respond varyingly to antiangiogenic therapy, making treatment less reliable than originally hoped (Folkman, 1972). The efficacy of antiangiogenic treatments is dependent on tumor stage. The angiogenesis inhibitor endostatin, as well as MMP and VEGF inhibitors, have been most successful in treating early-stage disease in the RIP-Tag mouse model, both by blocking growth and preventing the angiogenic switch. Other inhibitors are able to block proliferation and migration in end-stage tumors, but have no effect on early-stage disease prevention, suggesting that there may be qualitative differences in angiogenic vasculature or its regulation at different stages (reviewed in Bergers and Benjamin, 2003).

3. METASTASIS

Metastasis occurs when tumor cells leave the primary tumor and colonize another organ. For metastasis to occur, tumor cells disassociate from the primary tumor, invade and break down the extracellular matrix to enter the blood or lymphatic system, and disseminate to a new site, requiring survival in order to proliferate and colonize the new tissue (Chambers et al., 2000; Husemann and Klein, 2009; Weinberg, 2007). However, many questions remain, including what factors trigger metastatic spread, the genetic and epigenetic changes required, and the importance of selection or adaptation at the new site (Husemann and Klein, 2009).

Modeling metastasis has been difficult in the mouse, although mouse models have played an important role in studying the role of genetic background in metastasis, as discussed above (Section II.E.3). Spontaneous metastasis is rare in the mouse, and those tumors that do metastasize have long latency periods. Even for genetically engineered mouse models of metastasis, penetrance is variable, often much lower than primary tumor incidence, and the primary tumor may need to be surgically removed to allow the study of metastases. In the majority of models, latency is still over 3 months (summarized in Khanna and Hunter, 2005). However, mouse models of metastasis will be able to address specific questions that cannot be answered in cell culture or in human studies, such as what are the specific interactions that occur between the tumor and microenvironment that facilitate specific mechanisms of metastasis.

Recently, mouse models have provided important information regarding the metastatic potential of tumors, and the timing of acquiring this potential. It had been thought that metastatic capability develops as a late step in tumor progression, due to continuing accumulation of somatic mutations in the primary tumor and selection for metastatic ability in a small subset of cells (Fearon and Vogelstein, 1990; Fidler and Kripke, 1977). However, recent data suggests that tumors capable of metastasis actually show a specific and predictive gene expression fingerprint very early in tumorigenesis (Ramaswamy et al., 2003; van 't Veer et al., 2002). Similar findings have also been made in the mouse, as shown by Qiu et al. (2004). They demonstrate that a gene expression signature can distinguish between MMTV-PyMT mammary tumors in a highly metastatic strain and tumors in a low metastasis strain. Importantly, when looking specifically at orthologs of the 17 differentially expressed genes in the human fingerprint (Ramaswamy et al., 2003), 16 of the 17 genes show the same directional expression differences in the mouse correlating to a calculated metastatic potential (Qiu et al., 2004). These findings support the idea that a small number of metastasis-associated genes may be predictive in both humans and mouse. Interestingly, because these arrays were done comparing the signatures of primary tumors, these findings suggest that the metastatic predictive signature is not a rare phenomenon seen only in a small group of cells. Similarly, as the tumors in this experiment were all generated by the same oncogenic event, this suggests that the differences are most likely due to genetic background rather than oncogenic mutations. Therefore, qPCR was performed on normal mammary tissue from strains with high and low metastatic potential, using 10 of the 16 metastatic predictive genes, and it was found that nine of these signature genes show differential expression in normal tissues as well (Yang et al., 2005). These data suggest that gene expression patterns that predict metastatic ability may be influenced by hereditary polymorphisms, rather than just genetic mutations. It also suggests that certain subgroups of humans may be more susceptible to metastases than others, and that nontumor tissue can be used to predict risk (Qiu et al., 2004).

Other studies in the mouse have also supported the idea that metastasis is not a late step in tumor development. Another study of mammary tumors in MMTV rat *HER-2/neu* transgenic BALB/c mice found that HER-2+ mammary cells were detected in the bone marrow as early as 4–9 weeks, during a period of atypical ductal hyperplasia, but prior to carcinoma development (Husemann et al., 2008). Similarly, MMTV-PyMT mice had tumor cells present in the bone marrow at 4–6 weeks, with lung micrometastases found beginning at 14 weeks (Husemann et al., 2008). Electron microscopy confirmed that individual cells could be seen breaking through the basement membrane at early time points (Husemann et al., 2008).

As mentioned previously, genetically engineered mouse models are useful for imaging, as techniques for early detection become more accurate, and tumor cells can be labeled to follow cell growth. The ability to detect metastatic cell clusters and fully formed metastases by imaging can be used to improve the understanding of how metastases target specific organs. Recent studies have been done following the fate of single cells in mouse cancer models, and have given insight into interactions with the microenvironment, as well as the efficiency of cell survival in ectopic tissues (Chambers *et al.*, 2000; Khanna *et al.*, 2004).

4. PRECLINICAL TESTING OF THERAPEUTICS IN MOUSE MODELS

Though our understanding of cancer has expanded greatly in the past 40 years, our ability to translate that knowledge into patient treatment and therapy has been much more limited. Although genetically engineered mouse models have now been widely used to study and understand the basic science and molecular mechanisms of cancer, applying these models to preclinical studies of candidate therapeutics has been slower to develop. Most preclinical testing has been done in cell culture systems *in vitro* and xenografts *in vivo* due to the ability to control timing of tumorigenesis and generate large numbers of synchronized test subjects. Unfortunately, while these model systems have been successful at predicting toxicity issues in humans, they have been less successful at predicting efficacy of antitumor compounds. Genetically engineered mouse models are more difficult and costly to use for preclinical testing due to the number of animals that need to be bred to generate a test cohort; however, these models have certain advantages in modeling human disease that may make them more predictive of efficacy in humans. Because of the expense of clinical trials and the limited population of patients to be entered into clinical trials, the increased cost of using genetically engineered models may be offset by the increased ability to predict responses in humans, reducing the number of human trials that need to be performed. Genetically engineered mouse models of cancer have the advantages that the initiating mutation is known, which is particularly important for the testing of molecularly targeted therapies; that tumors develop spontaneously in the normal tissue for that tumor type, with a coevolving microenvironment; and that the immune system is intact (Becher and Holland, 2006). The advances in mouse models discussed in this chapter may provide the tools to better predict drug efficacy prior to Phase II/III drug trials.

Improvements in drug discovery and screening will require numerous steps to which improved mouse models may be able to contribute. These include the identification of cancer targets, including the context in which they are required for tumor maintenance, the determinations of most effective and least toxic compounds on these targets, and the identification of biomarkers (reviewed in Gutmann et al., 2006; Sharpless and Depinho, 2006). Genetically engineered mouse models can play an important role in target validation, because specific genetic events in human cancer are used to guide the development of the mouse model. Many genetically engineered tumor models have been generated to mimic the genetic, molecular, and phenotypic traits of the specific human cancer (Becher and Holland, 2006).

In order to make genetically engineered models effective for preclinical drug screenings, models with short tumor latency and high penetrance will help to keep maintenance manageable and costs reduced. However, an increase in mutations in order to decrease latency may skew tumor development, as well as reduce the requirement for gain of secondary or epigenetic mutations. A K-ras model of pancreatic cancer shows that in an *Ink4a/Arf* heterozygous background, the primary tumors have increased genomic complexity than those on an *Ink4a/Arf$^{-/-}$* background (Bardeesy et al., 2006), suggesting that a less biased model leads to greater genetic instability, as seen in human cancers, and is therefore a more accurate representation of the human tumor. Additionally, fast-growing tumors may not accurately reflect the interaction between the tumor cells and the microenvironment.

Genetically engineered mouse models can be used for systematic screening of compounds, and have been used to test drugs for both therapeutic and cancer prevention properties, as reviewed in Carver and Pandolfi (2006), Gutmann et al. (2006), and Sharpless and Depinho (2006). Potential therapies can be screened and compared when tested on a well-defined model. Effective drug therapy can also be stopped to determine if the drug cures the cancer, or if recurrence or drug resistance occurs. Multiple drug combinations can also be tested for synergistic effects in the mouse. Because recent molecularly targeted therapies are showing very specific efficacy in a subset of human patients (e.g., see Haas-Kogan et al., 2005; Mellinghoff et al., 2005), compounds can be tested on modeled tumors with different engineered mutations or on different strain backgrounds to determine whether specific efficacy can be predicted in advance to allow better design of clinical trials.

IV. SUMMARY

As reviewed here, advances in mouse modeling techniques continue to allow the further understanding of the biology of tumor development and growth. Mouse models have shown some of the complexities of tumor cell interactions, including tumor interaction with nonneoplastic cells in the microenvironment and the effect of normal genetic polymorphisms on tumor susceptibility. Identifying cancer-causing genes, and most specifically targets that are responsive to therapeutics, continues to be a challenge. New modalities of genetic engineering have given us the ability to more accurately model human cancer in an effort to identify new targets and markers that will allow us better predictive and treatment abilities. Future therapies will utilize the advances that have been made in these mice to further understand and treat cancer.

ACKNOWLEDGMENTS

The authors thank Shyam Sharan and Bernard Ramsahoye for helpful comments on the text.

REFERENCES

Adams, J. M., and Strasser, A. (2008). Is tumor growth sustained by rare cancer stem cells or dominant clones? *Cancer Res.* **68**, 4018–4021.
Adams, D. J., *et al.* (2004). Mutagenic insertion and chromosome engineering resource (MICER). *Nat. Genet.* **36**, 867–871.
Akagi, K., *et al.* (2004). RTCGD: Retroviral tagged cancer gene database. *Nucleic Acids Res.* **32**, D523–D527.
Allinen, M., *et al.* (2004). Molecular characterization of the tumor microenvironment in breast cancer. *Cancer Cell* **6**, 17–32.
Arbeit, J. M., *et al.* (1994). Progressive squamous epithelial neoplasia in K14-human papillomavirus type 16 transgenic mice. *J. Virol.* **68**, 4358–4368.
Artandi, S. E., *et al.* (2000). Telomere dysfunction promotes non-reciprocal translocations and epithelial cancers in mice. *Nature* **406**, 641–645.
Attardi, L. D., and Donehower, L. A. (2005). Probing p53 biological functions through the use of genetically engineered mouse models. *Mutat. Res.* **576**, 4–21.
Balkwill, F., and Coussens, L. M. (2004). Cancer: An inflammatory link. *Nature* **431**, 405–406.
Bardeesy, N., *et al.* (2006). Both p16(Ink4a) and the p19(Arf)-p53 pathway constrain progression of pancreatic adenocarcinoma in the mouse. *Proc. Natl. Acad. Sci. USA* **103**, 5947–5952.
Barlow, C., *et al.* (1996). Atm-deficient mice: A paradigm of ataxia telangiectasia. *Cell* **86**, 159–171.
Baron, U., *et al.* (1997). Tetracycline-controlled transcription in eukaryotes: Novel transactivators with graded transactivation potential. *Nucleic Acids Res.* **25**, 2723–2729.
Becher, O. J., and Holland, E. C. (2006). Genetically engineered models have advantages over xenografts for preclinical studies. *Cancer Res.* **66**, 3355–3358; discussion 3358–3359.
Ben David, Y., *et al.* (1988). Inactivation of the p53 oncogene by internal deletion or retroviral integration in erythroleukemic cell lines induced by Friend leukemia virus. *Oncogene* **3**, 179–185.
Berger, S. L., *et al.* (1990). Selective inhibition of activated but not basal transcription by the acidic activation domain of VP16: Evidence for transcriptional adaptors. *Cell* **61**, 1199–1208.

Bergers, G., and Benjamin, L. E. (2003). Tumorigenesis and the angiogenic switch. *Nat. Rev. Cancer* **3**, 401–410.

Bergers, G., *et al.* (2000). Matrix metalloproteinase-9 triggers the angiogenic switch during carcinogenesis. *Nat. Cell Biol.* **2**, 737–744.

Bibby, M. C. (2004). Orthotopic models of cancer for preclinical drug evaluation: Advantages and disadvantages. *Eur. J. Cancer* **40**, 852–857.

Bibikova, M., *et al.* (2006). Human embryonic stem cells have a unique epigenetic signature. *Genome Res.* **16**, 1075–1083.

Bockamp, E., *et al.* (2008). Conditional transgenic mouse models: From the basics to genome-wide sets of knockouts and current studies of tissue regeneration. *Regen. Med.* **3**, 217–235.

Branda, C. S., and Dymecki, S. M. (2004). Talking about a revolution: The impact of site-specific recombinases on genetic analyses in mice. *Dev. Cell* **6**, 7–28.

Broman, K. W. (2005). The genomes of recombinant inbred lines. *Genetics* **169**, 1133–1146.

Brown, S. D., *et al.* (2008). Quiet as a mouse: Dissecting the molecular and genetic basis of hearing. *Nat. Rev. Genet.* **9**, 277–290.

Buchholz, F., *et al.* (2000). Inducible chromosomal translocation of AML1 and ETO genes through Cre/loxP-mediated recombination in the mouse. *EMBO Rep.* **1**, 133–139.

Capecchi, M. R. (2005). Gene targeting in mice: Functional analysis of the mammalian genome for the twenty-first century. *Nat. Rev. Genet.* **6**, 507–512.

Carver, B. S., and Pandolfi, P. P. (2006). Mouse modeling in oncologic preclinical and translational research. *Clin. Cancer Res.* **12**, 5305–5311.

Chai, Y., *et al.* (2000). Fate of the mammalian cranial neural crest during tooth and mandibular morphogenesis. *Development* **127**, 1671–1679.

Chambers, A. F., *et al.* (2000). Molecular biology of breast cancer metastasis. Clinical implications of experimental studies on metastatic inefficiency. *Breast Cancer Res.* **2**, 400–407.

Chang, S., *et al.* (2009). Expression of human BRCA1 variants in mouse ES cells allows functional analysis of BRCA1 mutations. *J. Clin. Invest.* **119**, 3160–3171.

Chesler, E. J., *et al.* (2008). The Collaborative Cross at Oak Ridge National Laboratory: Developing a powerful resource for systems genetics. *Mamm. Genome* **19**, 382–389.

Chin, L., *et al.* (1999). p53 deficiency rescues the adverse effects of telomere loss and cooperates with telomere dysfunction to accelerate carcinogenesis. *Cell* **97**, 527–538.

Cho, B. C., *et al.* (1995). Frequent disruption of the Nf1 gene by a novel murine AIDS virus-related provirus in BXH-2 murine myeloid lymphomas. *J. Virol.* **69**, 7138–7146.

Churchill, G. A., *et al.* (2004). The Collaborative Cross, a community resource for the genetic analysis of complex traits. *Nat. Genet.* **36**, 1133–1137.

Cichowski, K., *et al.* (1999). Mouse models of tumor development in neurofibromatosis type 1. *Science* **286**, 2172–2176.

Collier, L. S., *et al.* (2005). Cancer gene discovery in solid tumours using transposon-based somatic mutagenesis in the mouse. *Nature* **436**, 272–276.

Collier, L. S., *et al.* (2009). Whole-body sleeping beauty mutagenesis can cause penetrant leukemia/lymphoma and rare high grade glioma without associated embryonic lethality. *Cancer Res.* **69**, 8429–8437.

Collins, E. C., *et al.* (2000). Inter-chromosomal recombination of Mll and Af9 genes mediated by cre-loxP in mouse development. *EMBO Rep.* **1**, 127–132.

Cormier, R. T., *et al.* (1997). Secretory phospholipase Pla2g2a confers resistance to intestinal tumorigenesis. *Nat. Genet.* **17**, 88–91.

Coussens, L. M., *et al.* (1996). Genetic predisposition and parameters of malignant progression in K14-HPV16 transgenic mice. *Am. J. Pathol.* **149**, 1899–1917.

Coussens, L. M., *et al.* (1999). Inflammatory mast cells up-regulate angiogenesis during squamous epithelial carcinogenesis. *Genes Dev.* **13**, 1382–1397.

Crabtree, J. S., *et al.* (2001). A mouse model of multiple endocrine neoplasia, type 1, develops multiple endocrine tumors. *Proc. Natl. Acad. Sci. USA* **98**, 1118–1123.

Crawford, N. P., *et al.* (2006). Germline polymorphisms in SIPA1 are associated with metastasis and other indicators of poor prognosis in breast cancer. *Breast Cancer Res.* **8**, R16.

Crawford, Y., and Ferrara, N. (2009). VEGF inhibition: Insights from preclinical and clinical studies. *Cell Tissue Res.* **335**, 261–269.

Cronin, C. A., *et al.* (2001). The lac operator–repressor system is functional in the mouse. *Genes Dev.* **15**, 1506–1517.

Cullen, B. R. (2004). Transcription and processing of human microRNA precursors. *Mol. Cell.* **16**, 861–865.

D'Angelo, M., *et al.* (2000). Angiogenesis in transgenic models of multistep carcinogenesis. *J. Neurooncol.* **50**, 89–98.

Demant, P. (2003). Cancer susceptibility in the mouse: Genetics, biology and implications for human cancer. *Nat. Rev. Genet.* **4**, 721–734.

De Palma, M., *et al.* (2003). In vivo targeting of tumor endothelial cells by systemic delivery of lentiviral vectors. *Hum. Gene Ther.* **14**, 1193–1206.

de Visser, K. E., and Coussens, L. M. (2006). The inflammatory tumor microenvironment and its impact on cancer development. *Contrib. Microbiol.* **13**, 118–137.

de Visser, K. E., *et al.* (2005). De novo carcinogenesis promoted by chronic inflammation is B lymphocyte dependent. *Cancer Cell* **7**, 411–423.

de Visser, K. E., *et al.* (2006). Paradoxical roles of the immune system during cancer development. *Nat. Rev. Cancer* **6**, 24–37.

Dickins, R. A., *et al.* (2007). Tissue-specific and reversible RNA interference in transgenic mice. *Nat. Genet.* **39**, 914–921.

Dietrich, W. F., *et al.* (1993). Genetic identification of Mom-1, a major modifier locus affecting Min-induced intestinal neoplasia in the mouse. *Cell* **75**, 631–639.

Ding, S., *et al.* (2005). Efficient transposition of the piggyBac (PB) transposon in mammalian cells and mice. *Cell* **122**, 473–483.

Dragani, T. A. (2003). 10 years of mouse cancer modifier loci: Human relevance. *Cancer Res.* **63**, 3011–3018.

Dragatsis, I., and Zeitlin, S. (2001). A method for the generation of conditional gene repair mutations in mice. *Nucleic Acids Res.* **29**, E10.

DuPage, M., *et al.* (2009). Conditional mouse lung cancer models using adenoviral or lentiviral delivery of Cre recombinase. *Nat. Protoc.* **4**, 1064–1072.

Dupuy, A. J., *et al.* (2005). Mammalian mutagenesis using a highly mobile somatic Sleeping Beauty transposon system. *Nature* **436**, 221–226.

Dupuy, A. J., *et al.* (2009). A modified sleeping beauty transposon system that can be used to model a wide variety of human cancers in mice. *Cancer Res.* **69**, 8150–8156.

Eden, A., *et al.* (2003). Chromosomal instability and tumors promoted by DNA hypomethylation. *Science* **300**, 455.

Engle, S. J., *et al.* (2002). Elimination of colon cancer in germ-free transforming growth factor beta 1-deficient mice. *Cancer Res.* **62**, 6362–6366.

Evers, B., and Jonkers, J. (2006). Mouse models of BRCA1 and BRCA2 deficiency: Past lessons, current understanding and future prospects. *Oncogene* **25**, 5885–5897.

Fearon, E., and Vogelstein, B. (1990). A genetic model for colorectal tumorigenesis. *Cell* **61**, 759–767.

Fidler, I. J., and Kripke, M. L. (1977). Metastasis results from preexisting variant cells within a malignant tumor. *Science* **197**, 893–895.

Flint, J., *et al.* (2005). Strategies for mapping and cloning quantitative trait genes in rodents. *Nat. Rev. Genet.* **6**, 271–286.

Folkman, J. (1972). Anti-angiogenesis: New concept for therapy of solid tumors. *Ann. Surg.* **175,** 409–416.

Forster, A., *et al.* (2003). Engineering de novo reciprocal chromosomal translocations associated with Mll to replicate primary events of human cancer. *Cancer Cell* **3,** 449–458.

Forster, A., *et al.* (2005a). Chromosomal translocation engineering to recapitulate primary events of human cancer. *Cold Spring Harb. Symp. Quant. Biol.* **70,** 275–282.

Forster, A., *et al.* (2005b). The invertor knock-in conditional chromosomal translocation mimic. *Nat. Methods* **2,** 27–30.

Frese, K. K., and Tuveson, D. A. (2007). Maximizing mouse cancer models. *Nat. Rev. Cancer* **7,** 645–658.

Friedel, R. H., *et al.* (2007). EUCOMM—The European conditional mouse mutagenesis program. *Brief Funct. Genomic Proteomic* **6,** 180–185.

Friedrich, G., and Soriano, P. (1991). Promoter traps in embryonic stem cells: A genetic screen to identify and mutate developmental genes in mice. *Genes Dev.* **5,** 1513–1523.

Fujii, S. (2008). Exploiting dendritic cells and natural killer T cells in immunotherapy against malignancies. *Trends Immunol.* **29,** 242–249.

Gaudet, F., *et al.* (2003). Induction of tumors in mice by genomic hypomethylation. *Science* **300,** 489–492.

Gerber, H. P., *et al.* (2007). Mice expressing a humanized form of VEGF-A may provide insights into the safety and efficacy of anti-VEGF antibodies. *Proc. Natl. Acad. Sci. USA* **104,** 3478–3483.

German, J. (1993). Bloom syndrome: A mendelian prototype of somatic mutational disease. *Medicine (Baltimore)* **72,** 393–406.

Giraudo, E., *et al.* (2004). An amino-bisphosphonate targets MMP-9-expressing macrophages and angiogenesis to impair cervical carcinogenesis. *J. Clin. Invest.* **114,** 623–633.

Gonzalez-Murillo, A., *et al.* (2008). Unaltered repopulation properties of mouse hematopoietic stem cells transduced with lentiviral vectors. *Blood* **112,** 3138–3147.

Goodlad, R. A., *et al.* (2006). Inhibiting vascular endothelial growth factor receptor-2 signaling reduces tumor burden in the ApcMin/+ mouse model of early intestinal cancer. *Carcinogenesis* **27,** 2133–2139.

Goss, K. H., *et al.* (2002). Enhanced tumor formation in mice heterozygous for Blm mutation. *Science* **297,** 2051–2053.

Gossen, M., and Bujard, H. (1992). Tight control of gene expression in mammalian cells by tetracycline-responsive promoters. *Proc. Natl. Acad. Sci. USA* **89,** 5547–5551.

Gossen, J. A., *et al.* (1989). Efficient rescue of integrated shuttle vectors from transgenic mice: A model for studying mutations in vivo. *Proc. Natl. Acad. Sci. USA* **86,** 7971–7975.

Gould, K. A., *et al.* (1996). Mom1 is a semi-dominant modifier of intestinal adenoma size and multiplicity in Min/+ mice. *Genetics* **144,** 1769–1776.

Greten, F. R., *et al.* (2004). IKKbeta links inflammation and tumorigenesis in a mouse model of colitis-associated cancer. *Cell* **118,** 285–296.

Grinspan, J. B., *et al.* (1996). Re-entry into the cell cycle is required for bFGF-induced oligodendroglial dedifferentiation and survival. *J. Neurosci. Res.* **46,** 456–464.

Gross, L. (1978). Viral etiology of cancer and leukemia: A look into the past, present and future—G.H.A. Clowes Memorial Lecture. *Cancer Res.* **38,** 485–493.

Gupta, P. B., *et al.* (2009). Cancer stem cells: Mirage or reality? *Nat. Med.* **15,** 1010–1012.

Gutmann, D. H., *et al.* (2006). Harnessing preclinical mouse models to inform human clinical cancer trials. *J. Clin. Invest.* **116,** 847–852.

Haas-Kogan, D. A., *et al.* (2005). Epidermal growth factor receptor, protein kinase B/Akt, and glioma response to erlotinib. *J. Natl. Cancer Inst.* **97,** 880–887.

Hager, J. H., and Hanahan, D. (1999). Tumor cells utilize multiple pathways to down-modulate apoptosis. Lessons from a mouse model of islet cell carcinogenesis. *Ann. NY Acad. Sci.* **887**, 150–163.

Hambardzumyan, D., *et al.* (2008). Glioma formation, cancer stem cells, and akt signaling. *Stem Cell Rev.* **4**, 203–210.

Hande, M. P. (2004). DNA repair factors and telomere-chromosome integrity in mammalian cells. *Cytogenet. Genome Res.* **104**, 116–122.

Hansen, G. M., *et al.* (2008). Large-scale gene trapping in C57BL/6N mouse embryonic stem cells. *Genome Res.* **18**, 1670–1679.

Hill, R., *et al.* (2005). Selective evolution of stromal mesenchyme with p53 loss in response to epithelial tumorigenesis. *Cell* **123**, 1001–1011.

Hirai, H., *et al.* (2001). Oncogenic mechanisms of Evi-1 protein. *Cancer Chemother. Pharmacol.* **48**(Suppl 1), S35–S40.

Hirasawa, R., *et al.* (2008). Maternal and zygotic Dnmt1 are necessary and sufficient for the maintenance of DNA methylation imprints during preimplantation development. *Genes Dev.* **22**, 1607–1616.

Hoffman, R. M. (2009). Imaging cancer dynamics in vivo at the tumor and cellular level with fluorescent proteins. *Clin. Exp. Metastasis* **26**, 345–355.

Hosoda, T., *et al.* (2009). Clonality of mouse and human cardiomyogenesis in vivo. *Proc. Natl. Acad. Sci. USA* **106**, 17169–17174.

Hunter, K. W., and Crawford, N. P. (2008). The future of mouse QTL mapping to diagnose disease in mice in the age of whole-genome association studies. *Annu. Rev. Genet.* **42**, 131–141.

Hunter, K. W., *et al.* (2001). Predisposition to efficient mammary tumor metastatic progression is linked to the breast cancer metastasis suppressor gene Brms1. *Cancer Res.* **61**, 8866–8872.

Huse, J. T., and Holland, E. C. (2009). Genetically engineered mouse models of brain cancer and the promise of preclinical testing. *Brain Pathol.* **19**, 132–143.

Husemann, Y., and Klein, C. A. (2009). The analysis of metastasis in transgenic mouse models. *Transgenic Res.* **18**, 1–5.

Husemann, Y., *et al.* (2008). Systemic spread is an early step in breast cancer. *Cancer Cell* **13**, 58–68.

Iraqi, F. A., *et al.* (2008). The Collaborative Cross, developing a resource for mammalian systems genetics: A status report of the Wellcome Trust cohort. *Mamm. Genome* **19**, 379–381.

Ivics, Z., *et al.* (1997). Molecular reconstruction of Sleeping Beauty, a Tc1-like transposon from fish, and its transposition in human cells. *Cell* **91**, 501–510.

Jackson, E. L., *et al.* (2001). Analysis of lung tumor initiation and progression using conditional expression of oncogenic K-ras. *Genes Dev.* **15**, 3243–3248.

Jackson, M., *et al.* (2004). Severe global DNA hypomethylation blocks differentiation and induces histone hyperacetylation in embryonic stem cells. *Mol. Cell. Biol.* **24**, 8862–8871.

Jemal, A., *et al.* (2008). Annual report to the nation on the status of cancer, 1975–2005, featuring trends in lung cancer, tobacco use, and tobacco control. *J. Natl. Cancer Inst.* **100**, 1672–1694.

Jiang, X., *et al.* (2000). Fate of the mammalian cardiac neural crest. *Development* **127**, 1607–1616.

Johnson, L., *et al.* (2001). Somatic activation of the K-ras oncogene causes early onset lung cancer in mice. *Nature* **410**, 1111–1116.

Jones, P. A., and Baylin, S. B. (2007). The epigenomics of cancer. *Cell* **128**, 683–692.

Jonkers, J., and Berns, A. (1996). Retroviral insertional mutagenesis as a strategy to identify cancer genes. *Biochim. Biophys. Acta* **1287**, 29–57.

Kado, S., *et al.* (2001). Intestinal microflora are necessary for development of spontaneous adenocarcinoma of the large intestine in T-cell receptor beta chain and p53 double-knockout mice. *Cancer Res.* **61**, 2395–2398.

Kamijo, T., *et al.* (1999). Tumor spectrum in ARF-deficient mice. *Cancer Res.* **59**, 2217–2222.

Kanai, Y. (2008). Alterations of DNA methylation and clinicopathological diversity of human cancers. *Pathol. Int.* **58**, 544–558.

Kang, J. H., and Chung, J. K. (2008). Molecular-genetic imaging based on reporter gene expression. *J. Nucl. Med.* **49**(Suppl 2), 164S–179S.

Karlseder, J., *et al.* (1999). p53- and ATM-dependent apoptosis induced by telomeres lacking TRF2. *Science* **283**, 1321–1325.

Keng, V. W., *et al.* (2009). A conditional transposon-based insertional mutagenesis screen for genes associated with mouse hepatocellular carcinoma. *Nat. Biotechnol.* **27**, 264–274.

Khanna, C., and Hunter, K. (2005). Modeling metastasis in vivo. *Carcinogenesis* **26**, 513–523.

Khanna, C., *et al.* (2004). The membrane-cytoskeleton linker ezrin is necessary for osteosarcoma metastasis. *Nat. Med.* **10**, 182–186.

Kim, C. F., *et al.* (2005). Mouse models of human non-small-cell lung cancer: Raising the bar. *Cold Spring Harb. Symp. Quant. Biol.* **70**, 241–250.

Kim, M., *et al.* (2006). Comparative oncogenomics identifies NEDD9 as a melanoma metastasis gene. *Cell* **125**, 1269–1281.

Kohler, S. W., *et al.* (1991). Analysis of spontaneous and induced mutations in transgenic mice using a lambda ZAP/lacI shuttle vector. *Environ. Mol. Mutagen.* **18**, 316–321.

Kool, J., and Berns, A. (2009). High-throughput insertional mutagenesis screens in mice to identify oncogenic networks. *Nat. Rev. Cancer* **9**, 389–399.

Korsisaari, N., *et al.* (2007). Inhibition of VEGF-A prevents the angiogenic switch and results in increased survival of Apc+/min mice. *Proc. Natl. Acad. Sci. USA* **104**, 10625–10630.

Korsisaari, N., *et al.* (2008). Blocking vascular endothelial growth factor-A inhibits the growth of pituitary adenomas and lowers serum prolactin level in a mouse model of multiple endocrine neoplasia type 1. *Clin. Cancer Res.* **14**, 249–258.

Kost-Alimova, M., and Imreh, S. (2007). Modeling non-random deletions in cancer. *Semin. Cancer Biol.* **17**, 19–30.

Kundu, J. K., and Surh, Y. J. (2008). Inflammation: Gearing the journey to cancer. *Mutat. Res.* **659**, 15–30.

Lakso, M., *et al.* (1992). Targeted oncogene activation by site-specific recombination in transgenic mice. *Proc. Natl. Acad. Sci. USA* **89**, 6232–6236.

Lambert, R. (2009). Diversity Outbred and Collaborative Cross mice to offer maximum allelic variation. *In*: Jax Notes, Vol. 514. p. 2. The Jackson Laboratory, Bar Harbor, ME.

Lang, G. A., *et al.* (2004). Gain of function of a p53 hot spot mutation in a mouse model of Li-Fraumeni syndrome. *Cell* **119**, 861–872.

Largaespada, D. A., *et al.* (1995). Retroviral insertion at the Evi-2 locus in BXH-2 myeloid leukemia cell lines disrupts Nf1 expression without changes in steady state ras-GTP levels. *J. Virol.* **69**, 5095–5102.

Lawlor, E. R., *et al.* (2006). Reversible kinetic analysis of Myc targets in vivo provides novel insights into Myc-mediated tumorigenesis. *Cancer Res.* **66**, 4591–4601.

Lee, H. W., *et al.* (1998). Essential role of mouse telomerase in highly proliferative organs. *Nature* **392**, 569–574.

Lee, E. M., *et al.* (2007). Xenograft models for the preclinical evaluation of new therapies in acute leukemia. *Leuk. Lymphoma* **48**, 659–668.

Lee, J., *et al.* (2008). Epigenetic-mediated dysfunction of the bone morphogenetic protein pathway inhibits differentiation of glioblastoma-initiating cells. *Cancer Cell* **13**, 69–80.

Legrand, N., *et al.* (2009). Humanized mice for modeling human infectious disease: Challenges, progress, and outlook. *Cell Host Microbe* **6**, 5–9.

Lei, H., *et al.* (1996). De novo DNA cytosine methyltransferase activities in mouse embryonic stem cells. *Development* **122**, 3195–3205.

LePage, D. F., and Conlon, R. A. (2006). Animal models for disease: Knockout, knock-in, and conditional mutant mice. *Methods Mol. Med.* **129**, 41–67.

Lessard, J., and Sauvageau, G. (2003). Bmi-1 determines the proliferative capacity of normal and leukaemic stem cells. *Nature* **423**, 255–260.

Lifsted, T., *et al.* (1998). Identification of inbred mouse strains harboring genetic modifiers of mammary tumor age of onset and metastatic progression. *Int. J. Cancer* **77**, 640–644.

Lin, E. Y., *et al.* (2001). Colony-stimulating factor 1 promotes progression of mammary tumors to malignancy. *J. Exp. Med.* **193**, 727–740.

Lipsick, J. S., and Wang, D. M. (1999). Transformation by v-Myb. *Oncogene* **18**, 3047–3055.

Liu, X., *et al.* (2007). Somatic loss of BRCA1 and p53 in mice induces mammary tumors with features of human BRCA1-mutated basal-like breast cancer. *Proc. Natl. Acad. Sci. USA* **104**, 12111–12116.

Lobato, M. N., *et al.* (2008). Modeling chromosomal translocations using conditional alleles to recapitulate initiating events in human leukemias. *J. Natl. Cancer Inst. Monogr.* **39**, 58–63.

Lozano, G., and Behringer, R. R. (2007). New mouse models of cancer: Single-cell knockouts. *Proc. Natl. Acad. Sci. USA* **104**, 4245–4246.

Luo, G., *et al.* (2000). Cancer predisposition caused by elevated mitotic recombination in Bloom mice. *Nat. Genet.* **26**, 424–429.

Lyons, S. K. (2005). Advances in imaging mouse tumour models in vivo. *J. Pathol.* **205**, 194–205.

Macleod, K. F., and Jacks, T. (1999). Insights into cancer from transgenic mouse models. *J. Pathol.* **187**, 43–60.

MacPhee, M., *et al.* (1995). The secretory phospholipase A2 gene is a candidate for the Mom1 locus, a major modifier of ApcMin-induced intestinal neoplasia. *Cell* **81**, 957–966.

Maddison, K., and Clarke, A. R. (2005). New approaches for modelling cancer mechanisms in the mouse. *J. Pathol.* **205**, 181–193.

Maedler, K., *et al.* (2002). Glucose-induced beta cell production of IL-1beta contributes to glucotoxicity in human pancreatic islets. *J. Clin. Invest.* **110**, 851–860.

Maggi, A., *et al.* (2004). Techniques: Reporter mice—A new way to look at drug action. *Trends Pharmacol. Sci.* **25**, 337–342.

Martinez, R., *et al.* (2009). A microarray-based DNA methylation study of glioblastoma multiforme. *Epigenetics* **4**, 255–264.

Marumoto, T., *et al.* (2009). Development of a novel mouse glioma model using lentiviral vectors. *Nat. Med.* **15**, 110–116.

Maser, R. S., *et al.* (2007). Chromosomally unstable mouse tumours have genomic alterations similar to diverse human cancers. *Nature* **447**, 966–971.

Meister, G., and Tuschl, T. (2004). Mechanisms of gene silencing by double-stranded RNA. *Nature* **431**, 343–349.

Mellinghoff, I. K., *et al.* (2005). Molecular determinants of the response of glioblastomas to EGFR kinase inhibitors. *N. Engl. J. Med.* **353**, 2012–2024.

Metzger, D., and Chambon, P. (2001). Site- and time-specific gene targeting in the mouse. *Methods* **24**, 71–80.

Mikkola, H. K., and Orkin, S. H. (2005). Gene targeting and transgenic strategies for the analysis of hematopoietic development in the mouse. *Methods Mol. Med.* **105**, 3–22.

Momota, H., and Holland, E. C. (2005). Bioluminescence technology for imaging cell proliferation. *Curr. Opin. Biotechnol.* **16**, 681–686.

Montini, E., *et al.* (2009). The genotoxic potential of retroviral vectors is strongly modulated by vector design and integration site selection in a mouse model of HSC gene therapy. *J. Clin. Invest.* **119**, 964–975.

Moon, J. H., *et al.* (2008). Induction of neural stem cell-like cells (NSCLCs) from mouse astrocytes by Bmi1. *Biochem. Biophys. Res. Commun.* **371**, 267–272.

Morahan, G., *et al.* (2008). Establishment of "The Gene Mine": A resource for rapid identification of complex trait genes. *Mamm. Genome* **19**, 390–393.

Morishita, K., *et al.* (1988). Retroviral activation of a novel gene encoding a zinc finger protein in IL-3-dependent myeloid leukemia cell lines. *Cell* **54**, 831–840.

Moser, A. R., *et al.* (1992). The Min (multiple intestinal neoplasia) mutation: Its effect on gut epithelial cell differentiation and interaction with a modifier system. *J. Cell Biol.* **116**, 1517–1526.

Mowat, M., *et al.* (1985). Rearrangements of the cellular p53 gene in erythroleukaemic cells transformed by Friend virus. *Nature* **314**, 633–636.

Muzumdar, M. D., *et al.* (2007). Modeling sporadic loss of heterozygosity in mice by using mosaic analysis with double markers (MADM). *Proc. Natl. Acad. Sci. USA* **104**, 4495–4500.

Nadeau, J. H., *et al.* (2000). Analysing complex genetic traits with chromosome substitution strains. *Nat. Genet.* **24**, 221–225.

Ngan, E. S., *et al.* (2002). The mifepristone-inducible gene regulatory system in mouse models of disease and gene therapy. *Semin. Cell Dev. Biol.* **13**, 143–149.

Niwa, H. (2007). How is pluripotency determined and maintained? *Development* **134**, 635–646.

No, D., *et al.* (1996). Ecdysone-inducible gene expression in mammalian cells and transgenic mice. *Proc. Natl. Acad. Sci. USA* **93**, 3346–3351.

Nohmi, T., *et al.* (2000). Recent advances in the protocols of transgenic mouse mutation assays. *Mutat. Res.* **455**, 191–215.

Normanno, N., *et al.* (2009). Target-based therapies in breast cancer: Current status and future perspectives. *Endocr. Relat. Cancer* **16**, 675–702.

Odelberg, S. J. (2002). Inducing cellular dedifferentiation: A potential method for enhancing endogenous regeneration in mammals. *Semin. Cell Dev. Biol.* **13**, 335–343.

O'Hagan, R. C., *et al.* (2002). Telomere dysfunction provokes regional amplification and deletion in cancer genomes. *Cancer Cell* **2**, 149–155.

Okano, M., *et al.* (1999). DNA methyltransferases Dnmt3a and Dnmt3b are essential for de novo methylation and mammalian development. *Cell* **99**, 247–257.

Olive, K. P., *et al.* (2009). Inhibition of Hedgehog signaling enhances delivery of chemotherapy in a mouse model of pancreatic cancer. *Science* **324**, 1457–1461.

Parangi, S., *et al.* (1996). Antiangiogenic therapy of transgenic mice impairs de novo tumor growth. *Proc. Natl. Acad. Sci. USA* **93**, 2002–2007.

Park, I. K., *et al.* (2003a). Bmi-1 is required for maintenance of adult self-renewing haematopoietic stem cells. *Nature* **423**, 302–305.

Park, Y. G., *et al.* (2003b). Multiple cross and inbred strain haplotype mapping of complex-trait candidate genes. *Genome Res.* **13**, 118–121.

Park, Y. G., *et al.* (2005). Sipa1 is a candidate for underlying the metastasis efficiency modifier locus Mtes1. *Nat. Genet.* **37**, 1055–1062.

Peeper, D., and Berns, A. (2006). Cross-species oncogenomics in cancer gene identification. *Cell* **125**, 1230–1233.

Pegram, M., and Ngo, D. (2006). Application and potential limitations of animal models utilized in the development of trastuzumab (Herceptin): A case study. *Adv. Drug Deliv. Rev.* **58**, 723–734.

Pelengaris, S., *et al.* (2002). Suppression of Myc-induced apoptosis in beta cells exposes multiple oncogenic properties of Myc and triggers carcinogenic progression. *Cell* **109**, 321–334.

Peters, L. L., *et al.* (2007). The mouse as a model for human biology: A resource guide for complex trait analysis. *Nat. Rev. Genet.* **8**, 58–69.

Pikarsky, E., *et al.* (2004). NF-kappaB functions as a tumour promoter in inflammation-associated cancer. *Nature* **431**, 461–466.

Porret, A., *et al.* (2006). Tissue-specific transgenic and knockout mice. *Methods Mol. Biol.* **337**, 185–205.

Pritchard, J. B., et al. (2003). The role of transgenic mouse models in carcinogen identification. *Environ. Health Perspect.* **111**, 444–454.

Prosser, H., and Bradley, A. (2003). Transgenics at breaking-point. *Cancer Cell* **3**, 411–413.

Qiu, T. H., et al. (2004). Global expression profiling identifies signatures of tumor virulence in MMTV-PyMT-transgenic mice: Correlation to human disease. *Cancer Res.* **64**, 5973–5981.

Qiu, W., et al. (2008). No evidence of clonal somatic genetic alterations in cancer-associated fibroblasts from human breast and ovarian carcinomas. *Nat. Genet.* **40**, 650–655.

Quintana, E., et al. (2008). Efficient tumour formation by single human melanoma cells. *Nature* **456**, 593–598.

Ramaswamy, S., et al. (2003). A molecular signature of metastasis in primary solid tumors. *Nat. Genet.* **33**, 49–54.

Reilly, K. M., et al. (2000). Nf1;Trp53 mutant mice develop glioblastoma with evidence of strain-specific effects. *Nat. Genet.* **26**, 109–113.

Reilly, K. M., et al. (2006). An imprinted locus epistatically influences Nstr1 and Nstr2 to control resistance to nerve sheath tumors in a neurofibromatosis type 1 mouse model. *Cancer Res.* **66**, 62–68.

Richmond, A., and Su, Y. (2008). Mouse xenograft models vs GEM models for human cancer therapeutics. *Dis. Model Mech.* **1**, 78–82.

Riggins, G. J., et al. (1995). Absence of secretory phospholipase A2 gene alterations in human colorectal cancer. *Cancer Res.* **55**, 5184–5186.

Sandy, P., et al. (2005). Mammalian RNAi: A practical guide. *Biotechniques.* **39**, 215–224.

Sausville, E. A., and Burger, A. M. (2006). Contributions of human tumor xenografts to anticancer drug development. *Cancer Res.* **66**, 3351–3354; discussion 3354.

Schnutgen, F., et al. (2005). Genomewide production of multipurpose alleles for the functional analysis of the mouse genome. *Proc. Natl. Acad. Sci. USA* **102**, 7221–7226.

Sharan, S. K., et al. (2009). Recombineering: A homologous recombination-based method of genetic engineering. *Nat. Protoc.* **4**, 206–223.

Sharpless, N. E., and Depinho, R. A. (2006). The mighty mouse: Genetically engineered mouse models in cancer drug development. *Nat. Rev. Drug Discov.* **5**, 741–754.

Shchors, K., and Evan, G. (2007). Tumor angiogenesis: Cause or consequence of cancer? *Cancer Res.* **67**, 7059–7061.

Shchors, K., et al. (2006). The Myc-dependent angiogenic switch in tumors is mediated by interleukin 1beta. *Genes Dev.* **20**, 2527–2538.

Shizuya, H., et al. (1992). Cloning and stable maintenance of 300-kilobase-pair fragments of human DNA in *Escherichia coli* using an F-factor-based vector. *Proc. Natl. Acad. Sci. USA* **89**, 8794–8797.

Shore, S. K., et al. (2002). Transforming pathways activated by the v-Abl tyrosine kinase. *Oncogene* **21**, 8568–8576.

Silva, J. M., et al. (2005). Second-generation shRNA libraries covering the mouse and human genomes. *Nat. Genet.* **37**, 1281–1288.

Singer, O., and Verma, I. M. (2008). Applications of lentiviral vectors for shRNA delivery and transgenesis. *Curr. Gene Ther.* **8**, 483–488.

Skarnes, W. C., et al. (2004). A public gene trap resource for mouse functional genomics. *Nat. Genet.* **36**, 543–544.

Soriano, P. (1999). Generalized lacZ expression with the ROSA26 Cre reporter strain. *Nat. Genet.* **21**, 70–71.

Soverini, S., et al. (2008). Imatinib mesylate for the treatment of chronic myeloid leukemia. *Expert Rev. Anticancer Ther.* **8**, 853–864.

Starr, T. K., et al. (2009). A transposon-based genetic screen in mice identifies genes altered in colorectal cancer. *Science* **323**, 1747–1750.

Stegmeier, F., *et al.* (2005). A lentiviral microRNA-based system for single-copy polymerase II-regulated RNA interference in mammalian cells. *Proc. Natl. Acad. Sci. USA* **102**, 13212–13217.

Steindler, D. A., and Laywell, E. D. (2003). Astrocytes as stem cells: Nomenclature, phenotype, and translation. *Glia* **43**, 62–69.

Stern, P., *et al.* (2008). A system for Cre-regulated RNA interference in vivo. *Proc. Natl. Acad. Sci. USA* **105**, 13895–13900.

Suzuki, T., *et al.* (2006). Tumor suppressor gene identification using retroviral insertional mutagenesis in Blm-deficient mice. *EMBO J.* **25**, 3422–3431.

Swing, D. A., and Sharan, S. K. (2004). BAC rescue: A tool for functional analysis of the mouse genome. *Methods Mol. Biol.* **256**, 183–198.

Takahashi, K., and Yamanaka, S. (2006). Induction of pluripotent stem cells from mouse embryonic and adult fibroblast cultures by defined factors. *Cell* **126**, 663–676.

Tan, B. T., *et al.* (2006). The cancer stem cell hypothesis: A work in progress. *Lab. Invest.* **86**, 1203–1207.

Tan, S. H., *et al.* (2008). Pharmacogenetics in breast cancer therapy. *Clin. Cancer Res.* **14**, 8027–8041.

Teicher, B. A. (2009). In vivo/ex vivo and in situ assays used in cancer research: A brief review. *Toxicol. Pathol.* **37**, 114–122.

Terabe, M., and Berzofsky, J. A. (2007). NKT cells in immunoregulation of tumor immunity: A new immunoregulatory axis. *Trends Immunol.* **28**, 491–496.

Terabe, M., and Berzofsky, J. A. (2008). The role of NKT cells in tumor immunity. *Adv. Cancer Res.* **101**, 277–348.

Theurillat, J. P., *et al.* (1999). Early induction of angiogenetic signals in gliomas of GFAP-v-src transgenic mice. *Am. J. Pathol.* **154**, 581–590.

Tiscornia, G., *et al.* (2004). CRE recombinase-inducible RNA interference mediated by lentiviral vectors. *Proc. Natl. Acad. Sci. USA* **101**, 7347–7351.

Tlsty, T. D., and Coussens, L. M. (2006). Tumor stroma and regulation of cancer development. *Annu. Rev. Pathol.* **1**, 119–150.

To, C., *et al.* (2004). The Centre for Modeling Human Disease Gene Trap resource. *Nucleic Acids Res.* **32**, D557–D559.

Tomlinson, I. P., *et al.* (1996). Variants at the secretory phospholipase A2 (PLA2G2A) locus: Analysis of associations with familial adenomatous polyposis and sporadic colorectal tumours. *Ann. Hum. Genet.* **60**, 369–376.

Troiani, T., *et al.* (2008). The use of xenograft models for the selection of cancer treatments with the EGFR as an example. *Crit. Rev. Oncol. Hematol.* **65**, 200–211.

Uren, A. G., *et al.* (2005). Retroviral insertional mutagenesis: Past, present and future. *Oncogene* **24**, 7656–7672.

Uren, A. G., *et al.* (2008). Large-scale mutagenesis in p19(ARF)- and p53-deficient mice identifies cancer genes and their collaborative networks. *Cell* **133**, 727–741.

Valdar, W., *et al.* (2006). Simulating the collaborative cross: Power of quantitative trait loci detection and mapping resolution in large sets of recombinant inbred strains of mice. *Genetics* **172**, 1783–1797.

van der Weyden, L., *et al.* (2009). Chromosome engineering in ES cells. *Methods Mol. Biol.* **530**, 49–77.

Van Dyke, T., and Jacks, T. (2002). Cancer modeling in the modern era: Progress and challenges. *Cell* **108**, 135–144.

van Lohuizen, M., and Berns, A. (1990). Tumorigenesis by slow-transforming retroviruses—An update. *Biochim. Biophys. Acta* **1032**, 213–235.

van Lohuizen, M., *et al.* (1989). N-myc is frequently activated by proviral insertion in MuLV-induced T cell lymphomas. *EMBO J.* **8**, 133–136.

van Lohuizen, M., et al. (1991). Identification of cooperating oncogenes in E mu-myc transgenic mice by provirus tagging. *Cell* **65**, 737–752.

van 't Veer, L. J., et al. (2002). Gene expression profiling predicts clinical outcome of breast cancer. *Nature* **415**, 530–536.

Ventura, A., et al. (2004). Cre-lox-regulated conditional RNA interference from transgenes. *Proc. Natl. Acad. Sci. USA* **101**, 10380–10385.

Walrath, J. C., et al. (2009). Chr 19(A/J) modifies tumor resistance in a sex- and parent-of-origin-specific manner. *Mamm. Genome* **20**, 214–223.

Wang, X., and Paigen, B. (2005). Genetics of variation in HDL cholesterol in humans and mice. *Circ. Res.* **96**, 27–42.

Wang, X. J., et al. (1999). Development of gene-switch transgenic mice that inducibly express transforming growth factor beta1 in the epidermis. *Proc. Natl. Acad. Sci. USA* **96**, 8483–8488.

Wang, X., et al. (2005). Identifying novel genes for atherosclerosis through mouse-human comparative genetics. *Am. J. Hum. Genet.* **77**, 1–15.

Wang, W., et al. (2007). Induced mitotic recombination of p53 in vivo. *Proc. Natl. Acad. Sci. USA* **104**, 4501–4505.

Wei, K., et al. (2002). Mouse models for human DNA mismatch-repair gene defects. *Trends Mol. Med.* **8**, 346–353.

Weinberg, R. A. (2007). Is metastasis predetermined? *Mol. Oncol.* **1**, 263–264; author reply 265–266.

Weinberg, R. A. (2008). Coevolution in the tumor microenvironment. *Nat. Genet.* **40**, 494–495.

Westbrook, T. F., et al. (2005). Dissecting cancer pathways and vulnerabilities with RNAi. *Cold Spring Harb. Symp. Quant. Biol.* **70**, 435–444.

Westphal, C. H., et al. (1997). atm and p53 cooperate in apoptosis and suppression of tumorigenesis, but not in resistance to acute radiation toxicity. *Nat. Genet.* **16**, 397–401.

Wilson, M. H., et al. (2007). PiggyBac transposon-mediated gene transfer in human cells. *Mol. Ther.* **15**, 139–145.

Yamamoto, M., et al. (2009). A multifunctional reporter mouse line for Cre- and FLP-dependent lineage analysis. *Genesis* **47**, 107–114.

Yang, H., et al. (2005). Metastasis predictive signature profiles pre-exist in normal tissues. *Clin. Exp. Metastasis* **22**, 593–603.

Yang, F. C., et al. (2008). Nf1-dependent tumors require a microenvironment containing Nf1 +/− and c-kit-dependent bone marrow. *Cell* **135**, 437–448.

Yu, L., et al. (2008). Interaction between bevacizumab and murine VEGF-A: A reassessment. *Invest. Ophthalmol. Vis. Sci.* **49**, 522–527.

Zardo, G., et al. (2008). Epigenetic plasticity of chromatin in embryonic and hematopoietic stem/progenitor cells: Therapeutic potential of cell reprogramming. *Leukemia* **22**, 1503–1518.

Zender, L., et al. (2006). Identification and validation of oncogenes in liver cancer using an integrative oncogenomic approach. *Cell* **125**, 1253–1267.

Zeng, Y., et al. (2002). Both natural and designed micro RNAs can inhibit the expression of cognate mRNAs when expressed in human cells. *Mol. Cell* **9**, 1327–1333.

Zheng, L., and Lee, W. H. (2002). Retinoblastoma tumor suppressor and genome stability. *Adv. Cancer Res.* **85**, 13–50.

Zheng, H., et al. (2008). Induction of abnormal proliferation by nonmyelinating Schwann cells triggers neurofibroma formation. *Cancer Cell* **13**, 117–128.

Zhu, Y., et al. (2002). Neurofibromas in NF1: Schwann cell origin and role of tumor environment. *Science* **296**, 920–922.

Zhu, Y., et al. (2005). Early inactivation of p53 tumor suppressor gene cooperating with NF1 loss induces malignant astrocytoma. *Cancer Cell* **8**, 119–130.

Index

A
Angiogenesis
  antiangiogenic treatment, 150
  astrocytoma, 149–150
  CXC chemokines
    bronchogenic cancer, 99–100
    gastrointestinal cancer, 98–99
    glioblastoma, 101
    head and neck cancer, 101
    melanoma, 97
    ovarian cancer, 98
    pancreatic cancer, 97–98
    prostate cancer, 100–101
    receptors, 93
    renal cell carcinoma, 101–102
  Myc activation, 149
  occurrence, 148
  vascular endothelial growth factor A (VEGF-A), 148–149
Angiostatic CXC chemokines
  ELR-negative, 95
  non-ELR, 95
  receptors
    CXCR4, 94–95
    G protein-coupled receptor (CXCR3), 94
Astrocytoma, 149–150

B
BK virus (BKV)
  agnoprotein, 6
  cancer, 10, 34
  clinical diseases, 10
  genome organization, 3, 6
  miRNAs expression, 6
  persistent infections, 9
  phylogenetic analysis, 3, 4
  seroprevalence rate, 7–8
  T antigen, 4
  transmission routes, 8
  viral life cycle, 7
Bronchogenic cancer, 99–100

C
Cancer. *See also* Metastasis
  cancer stem cell (CSC)
    conditional *Nf1* loss, 143–144
    epigenetic control, 144–145
    model, 141–142
    tumor-initiating cell, 142–143
  CXC chemokines
    angiostatic CXC chemokines, 94–95
    cancer angiogenesis, 92, 97–102
    cancer metastases, 92, 102–104
    immunoangiostasis, 95–96
  genetically engineered mouse (*see* Genetically engineered cancer mouse model)
  genetic changes, 114–115
  head and neck cancer, 101
  incidence and death, 114
  JC virus (JCV), 10, 34
  juvenile myelomonocytic leukemia, GOF myeloproliferative disorders, mice, 62
  N-SH2 domain residues, 60–61
  Karolinska Institute virus (KIV), 10
  Merkel cell polyomavirus (MCV)
    incidence, 10–11
    neuroendocrine derivation, 11
    pathological and immunophenotypic studies, 11
    PCR detection, 13
    polyomavirus large T antigen alignment, 14–16
    SV40 transformation potential, 13
    tumor-and nontumor-derived MCV, mutation, 12–13
    viral load determination, 11–12

Cancer. *See also* Metastasis (*continued*)
  molecular changes, 114
  ovarian cancer, 98
  prostate cancer, 100–101
  treatment, 115–116
  Washington University virus (WUV), 10
  xenograft model, 116
Choroid plexus epithelium (CPE), 32
Chronic inflammation
  premalignant tumor progression, 145–146
  signal pathway, 147
  tumor-promoting effect, 147–148
Clonal evolution model, 142
CNS development, Shp2
  adult brain, 72
  cerebellar development, 72
  ERK pathway activation, bFGF, 71–72
  mouse brain, 72
  neurogenic-to-gliogenic switch, 71
Conditional gene mutation, Cre/FLP recombinase
  adeno-and lentiviral vectors, 119
  gene expression and homologous recombination, 118–119
  tamoxifen-inducible system, 120
  tetracycline-repressor-based system, 119–120
Conditional overexpression model, 124–125
Constitutive transgenic model, 123–124
CXC chemokines
  angiostatic CXC chemokines
    ELR-negative, 95
    non-ELR, 95
    receptors, 94–95
  cancer angiogenesis, 92
    bronchogenic cancer, 99–100
    gastrointestinal cancer, 98–99
    glioblastoma, 101
    head and neck cancer, 101
    melanoma, 97
    ovarian cancer, 98
    pancreatic cancer, 97–98
    prostate cancer, 100–101
    renal cell carcinoma, 101–102
  cancer metastases, 92
    CXCL12-CXCR4 axis, 102–103
    CXCR7, 103–104
    process, 102
  immunoangiostasis, 95–96

D

Digital transcriptome subtraction (DTS), 2
DnaJ domain, 18–20
Duffy antigen for chemokines (DARC), 93

E

ELR-negative CXC chemokines, 95

G

Gain-of-function (GOF) mutation
  conditional overexpression model, 124–125
  constitutive transgenic model, 123–124
  mouse gene knockin model, 124
  Shp2
    juvenile myelomonocytic leukemia (JMML), 60–62
    Noonan syndrome, 55–56, 60
Gastrointestinal cancer, 98–99
Genetically engineered cancer mouse model
  angiogenesis
    antiangiogenic treatment, 150
    astrocytoma, 149–150
    Myc activation, 149
    occurrence, 148
    vascular endothelial growth factor A (VEGF-A), 148–149
  cancer
    genetic changes, 114–115
    incidence and death, 114
    molecular changes, 114
    stem cell and initiating cell, 141–145
    treatment, 115–116
    xenograft model, 116
  chromosomal translocation model
    invertor mice, 127
    translocator mice, 125–126
  chronic inflammation
    premalignant tumor progression, 145–146
    signal pathway, 147
    tumor-promoting effect, 147–148
  gene function
    conditional gene mutation, 118–120
    conditional overexpression model, 124–125
    constitutive transgenic model, 123–124
    gene knockin model, 124
    knockout gene, 117–118

RNA interference (RNAi), 120–121
single-cell knockout, 122–123
metastasis
cell clusters, 152
gene expression, 151
mechanism, 150
occurrence, 150
tumor progression, 151
therapeutical preclinical testing, 152–153
tumor genes
genetic modifier screens, 137–141
insertional mutagenesis, 132–134
transposons mutagenesis, 134–137
tumorigenesis mechanism and timing
genomic instability, 129–132
molecular-genetic imaging, 127–128
recombination reporters and lineage tracing, 128–129
Genetic modifier screens
classic mapping, 137–138
Collaborative Cross, 139–140
metastasis *(Mtes1)* modifier loci, 140
modifier of min (Mom) loci, 140
nerve sheath tumor resistance locus *(Nstr1)*, 140–141
reference strain panels, 137–139
Genomic instability
Big Blue® model, 131
Bloom syndrome, 130–131
driver and passenger mutations, 129–130
MutaTM model, 132
telomeres loss, 130
Germline PTPN11 mutations, 66–67
Glioblastoma, 101

H
Head and neck cancer, 101
Heart development, Shp2
cardiac malformations, 67–68
mouse model, 69–70
vertebrateS, 68

I
Immunoangiostasis, 95–96
Insertional mutagenesis
drawback, 134
mechanism, 133
mouse mammary tumor virus (MMTV), 133–134
murine leukemia virus (MuLV), 132–133
retroviruses, 132

J
JC virus (JCV)
cancer, 10, 34
genome organization, 3, 6
human neoplasia, 9
persistent infections, 9
phylogenetic analysis, 3, 4
reactivation, 9
seroprevalence, 7–8
transmission routes, 8
viral life cycle, 7
Juvenile myelomonocytic leukemia (JMML)
gain-of-function (GOF) mutation, Shp2
myeloproliferative disorders, mice, 62
N-SH2 domain residues, 60–61
PTPN11 mutations, 60–62

K
Karolinska Institute virus (KIV)
age specific prevalence studies, 7–8
agnoprotein, 6
cancer, 10
detection, 2
genome organization, 3, 6
phylogenetic analysis, 3, 4
transmission routes, 8
viral life cycle, 7
Knockout gene function
gene knockin model, 124
loss-of-function mouse model, 117–118
single-cell knockout, 122–123

L
Large T antigen (LT), SV40
Bub1, 21–22
Cul7, 20–21
DnaJ domain, 18–20
IRS1, 25–26
Nbs1, 26
p53, 26–28
pRB family, 23–25
LEOPARD syndrome
development, 67
germline PTPN11 mutations, 66–67
vs. Noonan syndrome, 67

Loss-of-function (LOF) mutation
  conditional gene mutations, 118–120
  mouse gene knockouts, 117–118
  RNA interference, 120–121
  Shp2
    invertebrates, 63–64
    LEOPARD syndrome, 66–67
    in mice, 55–56
    null mutations, mice, 64–66
    in Xenopus, 63–64
  single-cell knockout, 122–123

M
Mammary fat pad, 77
Melanoma, 97
Merkel cell polyomavirus (MCV)
  incidence, 10–11
  neuroendocrine derivation, 11
  pathological and immunophenotypic studies, 11
  PCR detection, 13
  polyomavirus large T antigen alignment, 14–16
  SV40 transformation potential, 13
  tumor-and nontumor-derived MCV, mutation, 12–13
  viral load determination, 11–12
Metastasis
  cell clusters, 152
  CXC chemokines, 92
    CXCL12-CXCR4 axis, 102–103
    CXCR7, 103–104
    process, 102
  gene expression, 151
  mechanism, 150
  occurrence, 150
  tumor progression, 151
Mosaic analysis with double markers (MADM), 123
Murine polyomavirus (MPyV)
  MCC, 16
  miRNAs expression, 6
  phylogenetic analysis, 3, 4
  PP2A, 31
  pRB protein, 23
Myeloid neoplasms, PTPN11 mutations, 62

N
Neural crest-derived tissues, Shp2
  cranial paraxial mesoderm, 75
  heart outflow tract and craniofacial structures, 74–75
  Schwann cells, 73–74
Neurogenic-to-gliogenic switch, 71
Non-ELR CXC chemokine, 95
Noonan syndrome
  gain-of-function (GOF) mutation, Shp2
    ERK MAP kinase activation, 60
    flies, 60
    mice, 55–56, 60
  symptoms, 59

O
Ovarian cancer, 98

P
p53 protein, 26–28
Pancreas development, Shp2, 75–76
Pancreatic cancer, 97–98
Phosphotyrosine phosphatase (PTP) domain, 54, 57
Polyomaviruses
  age specific prevalence studies, 7–8
  BK virus (see BK virus (BKV))
  classification, 3
  clinical disease, 9–10
  genome organization
  infection, 8–9
  JC virus (see JC virus (JCV))
  Karolinska Institute virus (see Karolinska Institute virus (KIV))
  merkel cell polyomavirus (see Merkel cell polyomavirus (MCV))
  murine polyomavirus (see Murine polyomavirus (MPyV))
  phylogeny, 3, 4
  SV40 (see Simian vacuolating virus 40 (SV40))
  T antigens
    alignment, 14–15
    splicing arrangement, 6
  viral life cycle, 7
  Washington University virus (see Washington University virus (WUV))
pRB family
  E1A, 25
  E2Fs, 24
  LT binding, LxCxE motif, 24–25
  p107 and p130, 23–24

pocket domain, 23
retinoblastic tumor, 23
Prostate cancer, 100–101
PTPN11 mutations
  JMML, 60–62
  myeloid neoplasms, 62
  solid tumors, 62–63

R

Renal cell carcinoma, 101–102
RNA-induced silencing complex (RISC), 120
RNA interference (RNAi), 120–121

S

Schwann cells, 73–74
Short hairpin RNA (shRNA), 120–121
Shp2-regulated signaling pathways, 58
Simian vacuolating virus 40 (SV40)
  genome organization, 5
  large T antigen (LT)
    Bub1, 21–22
    Cul7, 20–21
    DnaJ domain, 18–20
    IRS1, 25–26
    Nbs1, 26
    p53, 26–28
    pRB family, 23–25
  small T antigen (ST)
    DnaJ domain, 30
    growth, 29
    human cell transformation assay, 30–31
    phosphatidylinositol 3-kinase (PI3K) pathway, 31
    proliferation induction, 31–32
    protein phosphatase PP2A activity, 30
  transgenic expression
    choroid plexus epithelium (CPE), 32
    enterocytes LT and LT1–136, 33
    tumorigenesis, 33
    tumor induction, LT1–121, 32–33
  viral replication, 16–17
Sleeping Beauty transposon
  conditional system, 136–137
  somatic mutagenesis, 135–136
Small interfering RNA (siRNA), 120
Small T antigen (ST), SV40
  DnaJ domain, 30
  growth, 29
  human cell transformation assay, 30–31
  phosphatidylinositol 3-kinase (PI3K) pathway, 31
  proliferation induction, 31–32
  protein phosphatase PP2A activity, 30
Solid tumors, PTPN11 mutations, 62–63

T

T-cell development, Shp2, 70
Transposons mutagenesis, 134–137
Tyrosine phosphatase Shp2
  CNS
    adult brain, 72
    cerebellar development, 72
    ERK pathway activation, bFGF, 71–72
    mouse brain, 72
    neurogenic-to-gliogenic switch, 71
  gain-of-function (GOF) mutation
    cancer, 60–63
    Noonan syndrome, 59–61
    PTPN11 mutation, JMML, 60–63
  heart development
    cardiac malformations, 67–68
    mouse model, 69–70
    vertebrateS, 68
  liver, 76
  loss-of-function (LOF)
    invertebrates, 63–64
    LEOPARD syndrome, 66–67
    null mutations, mice, 64–66
    in Xenopus, 63–64
  mammary gland, 77
  neural crest-derived tissues
    cranial paraxial mesoderm, 75
    heart outflow tract and craniofacial structures, 74–75
    Schwann cells, 73–74
  pancreas, 75–76
  signaling pathways
    JAK/Stat pathway, 58–59
    phosphatidyl-inositol-3 kinase (PI3K)/Akt pathway, 59
    Ras/ERK MAP kinase pathway, 58
  structure, 54, 57–58
  T-cell development, 70

W

Washington University virus (WUV)
  age specific prevalence studies, 7–8
  cancer, 10

Washington University virus (WUV) (*continued*)
  detection, 2
  genome organization, 3, 6
  phylogenetic analysis, 3, 4
  transmission routes, 8
  viral life cycle, 7

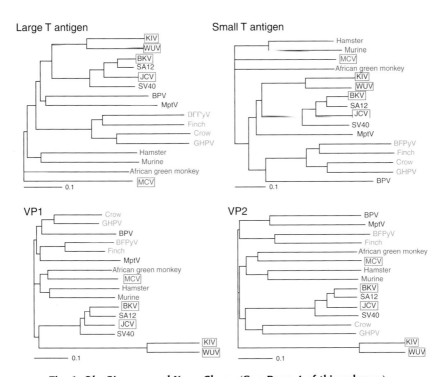

Fig. 1, Ole Gjoerup and Yuan Chang (See Page 4 of this volume.)

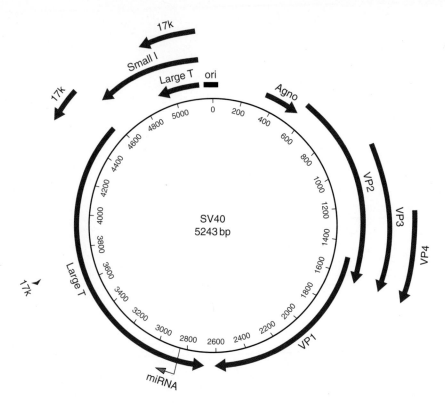

Fig. 2, Ole Gjoerup and Yuan Chang (See Page 5 of this volume.)

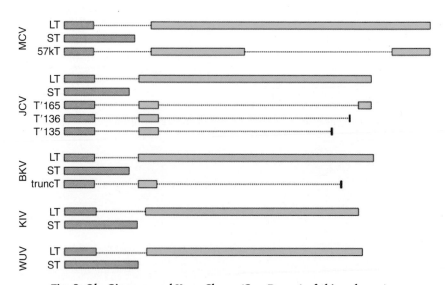

Fig. 3, Ole Gjoerup and Yuan Chang (See Page 6 of this volume.)

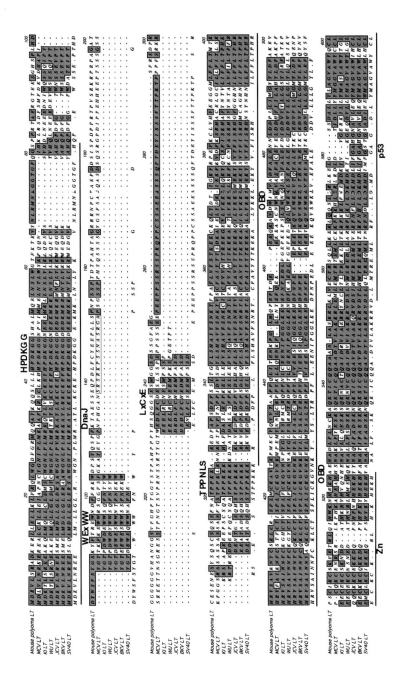

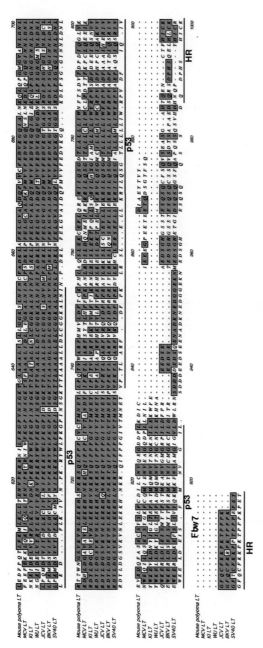

Fig. 4, Ole Gjoerup and Yuan Chang (See Page 15 of this volume.)